SMALL PETROL ENGINES

Operation and Maintenance

BRUCE HOLT

CW01498839

INKATA PRESS

INKATA PRESS

A DIVISION OF BUTTERWORTH-HEINEMANN

AUSTRALIA

BUTTERWORTH-HEINEMANN

North Tower 1–5 Railway Street

Chatswood NSW 2067

BUTTERWORTH-HEINEMANN

18 Salmon Street

Port Melbourne 3207

UNITED KINGDOM

BUTTERWORTH-HEINEMANN LTD

Oxford

USA

BUTTERWORTH-HEINEMANN

Stoneham

National Library of Australia Cataloguing-in-Publication entry

Holt, Bruce, 1947–
   Small petrol engines: operation and maintenance.

   Includes index.
   ISBN 0 7506 8901 3.

   1. Internal combustion engines, spark ignition – Maintenance
   and repair. I. Title. (Series: Practical farming).

621.434

Published by Reed International Books Australia. Under the Copyright Act 1968 (Cth), no part of this publication may be reproduced by any process, electronic or otherwise, without the specific written permission of the copyright owner.

Enquiries should be addressed to the publishers.

Typeset by Ian MacArthur, Hornsby Heights, NSW.

Printed in Australia by Ligare Pty Ltd, Riverwood, NSW.

SMALL PETROL ENGINES

To my children
Susan, Timothy and Emily

Contents

Preface

The small petrol engine is one of the most useful power sources within the rural industries of Australia. It has numerous applications, such as powering water pumps, grain augers, elevators, generators, compressors, motorbikes, mowers, concrete mixers and chainsaws.

To be able to efficiently operate, maintain and troubleshoot the small petrol engine, you need to have a basic understanding of the engine's operating principles and the way it is constructed. This book covers the basic operating principles and construction. It is to complement the operating and maintenance instructions that are given in the owner's manuals which the engine manufacturers dispatch with the new engine.

Chapter 1

Introduction

The small petrol engine belongs to the class of internal combustion engines of either the two or four stroke working cycle. The majority of engines used in rural Australia are of the four stroke variety and that is what this book mainly deals with.

The petrol engine is classified as a heat engine, for when petrol is burned, heat energy is released. The engine converts this heat energy into mechanical energy to do work.

There are two main types of heat engines:
- Internal combustion engines – like the small petrol engine where the petrol is combusted (burned) inside the engine cylinder.
- External combustion engines – like the piston steam engine where the fuel is combusted (burned) outside the engine cylinder.

Internal combustion is the act of burning inside an enclosed cylinder.

Internal combustion

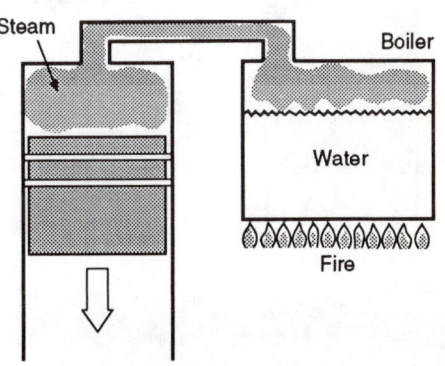

External combustion

To assist in understanding the petrol engine we need to go back in time to look at some of the history of engines.

For thousands of years draught animals such as horses provided the power to do most of the work that was required on the land. With the development of engines their power output was compared to the power of the horse and the words "horse power" came into use to give people an indication of the power performance of an engine.

The history of the development of the internal combustion engine goes back around two hundred years, with a fair degree of the development taking place in the nineteenth century.

Industry was using a workable four stroke engine in the late 1880s that ran mainly on coal gas. Others ran on a light oil.

The early engine developers used the mechanical principles of the piston steam engine of the eighteenth century to help them develop the internal combustion engine.

The ability of the steam engine to rotate a shaft powered the Industrial Revolution of the eighteenth and nineteenth centuries.

Because the internal combustion engine uses the same basic parts as the steam engine we need to look briefly and simply at the workings of steam to help us in our understanding of the workings of the internal combustion engine.

As was mentioned earlier, the fuel is burned externally in the steam engine. The fuel, such as coal or wood, burns to release heat so that the water in the boiler is heated and generates

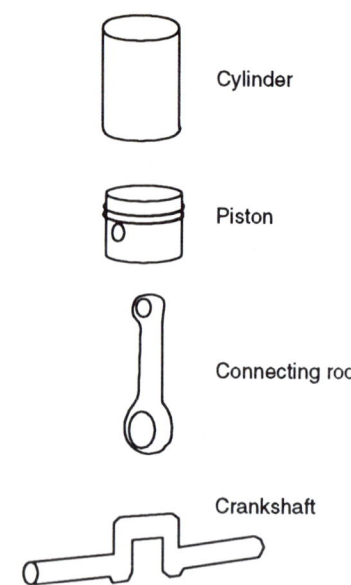

Cylinder

Piston

Connecting rod

Crankshaft

The piston steam engine uses a piston travelling in a cylinder to rotate a crankshaft to which it is linked by a connecting rod.

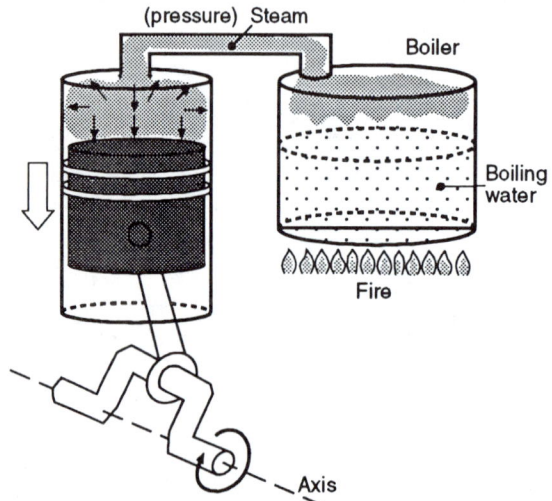

The diagram of the piston steam engine is simple and does not show all the working mechanisms.

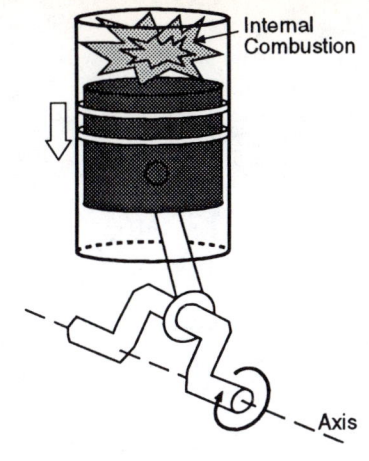

Internal
Combustion

Axis

Simple internal combustion engine

steam. The steam provides the pressure to act on the surface area of the piston and forces it down the cylinder, so that the crankshaft is rotated and work is done.

In the eighteenth century steam engines, only about 5 to 10 per cent of the heat energy of the fuel produced work at the crankshaft. The rest was lost.

Engine developers of the time began thinking how they would be able to force the piston down with a pressure other than that obtained from steam. Ideas ranged from burning gunpowder to coal gas. Gunpowder was ruled out because it burned violently and engines exploded.

Many people were involved in the development of the internal combustion engine, with a major breakthrough in 1876 when Dr Otto produced a workable four stroke engine that ran on coal gas. The same principles are still used today.

The two stroke engine was developed by such people as Clark in 1881 and refined in 1891 by Day.

In 1883 Daimler developed a small engine that ran on a light oil. It ran at 800 rpm and weighed 88 pounds per horse power.

Previously, engines ran on coal gas or heavy oils and ran at 200 rpm and weighed up to 1100 pounds per horse power.

The internal combustion engine provided a different form of rotating power that could be more widely used than the steam engine and also consumed less fuel for the power it produced.

One hundred years ago the two and four stroke engines were up and running. Since then the engines have been refined to make them more efficient, cheaper, lighter, smaller and easier to operate and maintain.

The older style engines were big, with their many moving parts exposed. Clearances and adjustments had liberal tolerances. Today's engines are small with most moving parts enclosed and clearances and adjustments have smaller tolerances. Today's engines require more care and attention to detail when it comes to their maintenance and repair.

Chapter 2

Safety

Safety is always on people's minds as they go about their daily lives, particularly in the work environment. Safety is about using common sense, caution and care.

The modern small petrol engine is a neat, compact engine as compared to the big exposed-flywheel engines of times past, but it still has to be treated with respect. When a new engine is purchased, an owner's manual is supplied. It contains operating and maintenance instructions that the engine manufacturer would like the owner of the engine to know about.

Usually at the start of the manual are the safety instructions, and advice from the manufacturer that the safety points should be read and understood and the manual fully read and understood before attempting to start and use the engine.

If the engine operates some type of equipment, commonly called "powered equipment", there is also an owner's manual supplied to cover the equipment. Some of this equipment can be quite dangerous, for example, the chainsaw. A main safety issue with using a chainsaw is the danger of "kickback" and this issue is well documented in a chainsaw owner's manual.

We shall now look at some of the safety issues that are associated with small petrol engines, naturally there are others but we will concentrate on the main ones.

Owner's Manual

Read it thoroughly and understand what is in it before operating the engine and any powered equipment.

People and Animals

People and animals should be kept away from operating

engines, and also from stopped engines that are still hot. Animals also have a habit of chewing on wiring, hoses and so on.

Protective Clothing

When operating or working on a small petrol engine, you should wear the appropriate protective clothing. A good chainsaw operator, for example, does not go out to work dressed in shorts and thongs.

A safe operator should have:
- Neat fitting, fire resistant clothing for protection from spilt petrol and hot or moving parts.
- Leather boots to protect feet.
- Eye protection when using compressed air or when foreign material is being thrown around by equipment such as a chainsaw.
- Ear protection when the noise level is high.
- Gloves to protect the hands from sharp parts like rewind springs.
- A hard hat for situations such as when using a chainsaw to fell trees.

Tidiness

The area around where the engine operates or is maintained in your workshop should be kept neat and tidy.

Tools

Always use the correct tool for the job. Adjustable spanners are not for using on engines.

Caution must be taken when using electrical devices, such as drills, in wet conditions to avoid short circuits or electrocution.

Lifting

When lifting an engine, do it properly. If it is light enough, use your thigh muscles and keep your back straight.

If it is a bit heavy get someone to help you or use an approved lifting device.

Compressed Air

When using compressed air to clean or maintain the engine, wear suitable clothing and face or eye protection. It is hard to close your eyes in time if the compressed air sends the dirt back at your face.

Petrol

Petrol is a liquid fuel that vaporises and ignites easily.

Petrol should be used within three months of being refined or it will go stale. Stale petrol does not work very well.

When refuelling do not fill the tank right up – leave an air gap.

Refuel outside or in a well ventilated area. Keep naked flames away from petrol. Do not smoke around engines or when refuelling.

If petrol is spilt, clean it up immediately and do not start the engine until spilt petrol has evaporated.

Move petrol containers away from the engine before starting the engine.

Store petrol in suitable containers. Store petrol containers in a safe dry location.

Do not refuel an engine that is running. Allow the engine to cool for at least two minutes before refuelling.

Engines should not be transported with fuel in the carburettors or tanks.

Fuel taps should be turned off if the engine is not in use.

Fire Risk

The area where the engine is operated should be cleared to reduce the risk of fire. Do not place rags on the engine. Clean engine as required to remove a build up of dry vegetation. Check fuel system and hoses for leaks.

Do not run the engine without an air cleaner or muffler. Use a spark arrester if the situation dictates it. Clean the muffler and spark arrester as recommended.

Exhaust Gases

Exhaust gases contain deadly carbon monoxide. Avoid breathing in the exhaust gases and run engines in well ventilated areas.

Powered Equipment

Always be cautious of powered equipment such as augers that use belts and pulleys. Keep your hands and feet clear of any moving parts on the equipment or the engines. Keep all guards in place.

Hot Parts

The cooling fins on a small engine do not change their colour in a hurry. So a cold engine can look the same as a hot engine but there is a difference of about 150° C. Be careful around the muffler and cooling fins.

Pre-start Checks

Before starting an engine, make sure you are familiar with the engine and its controls. Always check the engine to make sure it is ready to operate, with all the bits and pieces where they should be and oil and fuel levels checked.

Starting the Engine

Make sure you know how to stop the engine quickly if you have to. When pulling on a rewind starter or a hand wound rope, position the engine just past the top of the cycle of the compression stroke and then give it a good pull so as to get up the momentum to reduce the kickback effect. Most engines today use some type of compression release to make the engines easy to start.

Operating the Engine

Four stroke engines must run fairly level so that the lubrication and fuel systems work properly. Do not tamper with the governor, or overspeed or overload the engine. Always operate the engine with all parts in place.

Stopping the Engine

Before stopping the engine, remove the load (if possible) and reduce the engine speed to idle (if possible) to let the engine cool for a few minutes before stopping it. Do not use the choke to stop it.

Checking for Spark

When checking for spark, use a special tester. If you remove the plug and spin the flywheel, the petrol fumes that come out the plug hole could ignite and cause a fire or explosion. The spark plug hole should be fitted with a plug if you are creating sparks outside of the engine.

Routine Maintenance

Maintenance is carried out so that the engine will keep operating with the minimum amount of trouble to give long and faithful service. By carrying out routine maintenance you also clean the engine and check it over to make sure it is safe to keep operating it.

When carrying out work on the engine, disable the ignition to prevent accidental starting. Always read the owner's manual and follow the engine manufacturer's recommendations for maintenance and safety precautions.

Chapter 3

Hand Tools and Fasteners

Hand Tools

When you do maintenance work and light repairs to small petrol engines, you only need a small range of hand tools. The number of tools needed may be few but they must be the right tools for the job and they must be used correctly.

Spanner Sizes

The majority of small petrol engines currently used in rural Australia originated from Japan or the USA. Some earlier engines came from England. As a result of the engines coming from different countries, you need three different size ranges of spanners to suit the British, American and metric systems.

Early British engines used Whitworth bolts and nuts. The size of the Whitworth spanner is determined by the outer diameter of the shank of the bolt at the thread section. It may have both the BS and W markings on it.

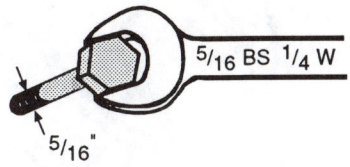

Early Australian farm machinery used Whitworth bolts and nuts so there are still Whitworth spanners around.

Engines that originate in the USA use mainly American threads, as do later British engines.

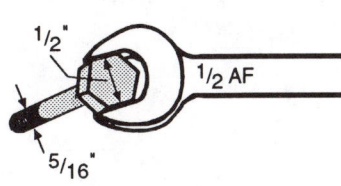

The size of a spanner to suit the American threads is determined by the distance across the flats of the nut or bolt it is to fit. The spanner may just have an imperial fraction size on it or it may also have AF (across flats) on it.

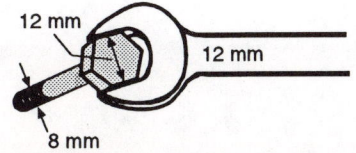

Engines that originate in Japan and most other overseas countries are metric. The size of a spanner to suit the metric threads is determined by the distance across the flats of the nut or bolt it is to fit. The spanner may just have a number on it (size in millimetres) or it may have the millimetres symbol (mm) also.

The three spanner sizes look about the same for a given bolt size but it is that little bit of difference that can cause damage to the bolt or nut if you use the wrong size spanner.

If you have a late model engine and you go to work on it with Whitworth spanners, not only are you going to take the skin off your knuckles when it slips off the bolt or nut, you are also going to round off the tips, so that the right spanner will possibly not fit correctly.

Before you go to work on an engine, establish what size spanners will fit it and that the spanners are clean and in good condition.

Poor quality tools should not be used unless you want poor results.

Loosening or Tightening a Bolt or Nut

If you are going to loosen or tighten a bolt or nut you should use a good fitting socket or ring spanner.

You should always pull on the spanner rather than push on it, so that you have better control over it and less chance of injury.

A hexagonal head on a bolt or nut has six sides. A double hex (twelve points) socket or ring spanner contacts the six tips. A single hex (six points) socket or ring spanner gives better contact and should be used if the fastener is damaged or very tight. A double hex spanner is the normal type to give better flexibility of use.

Using an open ended spanner to tighten or loosen a fastener can result in the tips being rounded. As force is applied to the spanner, it only contacts two flats, the jaws tend to flex and the spanner can slip around the tip.

Open ended spanners are designed to be used on fasteners that are loose.

The three types of spanners illustrated are all suitable for use on engines. If you are going to purchase some spanners the combination type is a good dual purpose tool: the ring end to loosen or tighten and the open end to use when the fastener is loose.

Double hex socket or ringspanner fitted to a hexagonal bolt.

Single hex socket or ringspanner fitted to a hexagonal bolt.

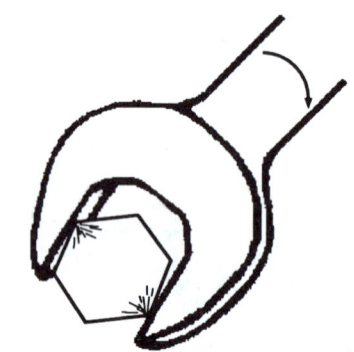

Using an open ended spanner to loosen or tighten a fastner is bad news.

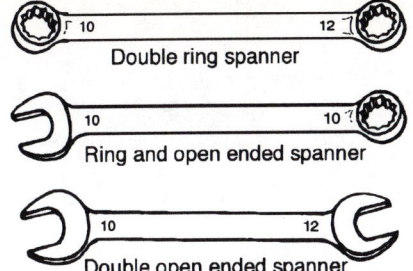

Double ring spanner

Ring and open ended spanner

Double open ended spanner

Sockets and Accessories

Sockets are fast and safe to use.

A 3/8″ drive set with both metric and imperial size sockets is ideal for working on most small engines.

If loosening or tightening a fastener, do not use the ratchet for it can be overloaded and damaged.

Deep sockets are also available for hard to get at fasteners.

Tube Spanners

Tube spanners are a long thin version of a socket and are suitable for use on fasteners that are not too tight. Do not overexert a tube spanner for it will distort and the fastener could be damaged.

A common use for a tube spanner is to remove and replace the spark plug.

Adjustable Spanners

Adjustable spanners have their place, but their use on small petrol engines should be avoided at all costs.

Torque Wrench

This is also referred to as a 'tension wrench'.

This wrench is used with a socket to tighten a fastener to a pre-determined limit. It is essential to ensure a bolt or nut is tensioned correctly to avoid warping or distorting components such as the head or connecting rod cap. It is a valuable tool to ensure fasteners are done up correctly, it eliminates the guesswork and reduces damage such as broken or stripped fasteners.

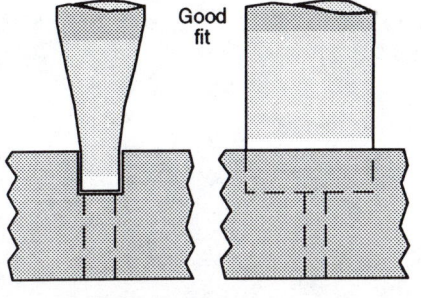

Good fit

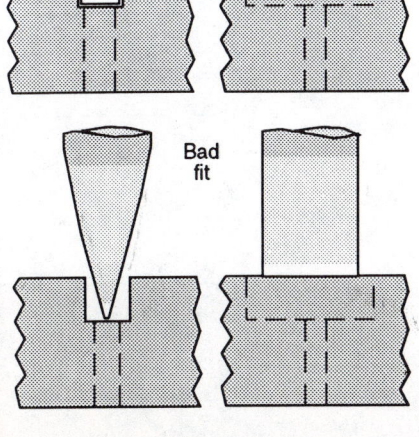

Bad fit

Screwdrivers

Two basic screwdrivers are used on small engines:
• the conventional flat blade
• Phillips head.

The screwdriver is a very good basic tool, but if not properly used can cause damage in a matter of seconds.

The most common area of damage from screwdrivers that do not fit the slot correctly is in the removal and replacement of the jets in the carburettor. The jets are usually brass and are easily damaged.

Feeler Gauges

Feeler gauges are used for measuring small gaps such as at the spark plug, contact points, armature air gap and tappet clearance. The gap is usually stated in fractions of a millimetre or thousandths of an inch.

The feeler gauges come in two types:
* blade type
* wire type.

The blade type is the general purpose type for measuring gaps at the points or armature or tappet.

The wire type is used for measuring the gap between the electrodes on spark plugs.

Special Tools

Special tools, such as flywheel pullers and holders, are available from engine dealers, if you are going to perform service or repair work on a engine.

Fasteners

'Fastener' is the term used to describe a device that holds two or more parts or pieces together.

The main fasteners in a small engine that we are concerned with are nuts, bolts and screws.

Bolt and Nut

A bolt is a metal rod that has a head, usually six sided, on one end and a thread for a nut on the other end. The main job of the bolt is to hold two or more parts together with the aid of a nut screwed onto the threaded end and tightened.

A nut is a piece of metal, usually six sided with a threaded hole so it can be screwed onto a bolt. Nuts can be of many types, such as plain, self locking or slotted.

Cap Screw

A cap screw, or machine screw, is a fastener that screws into a threaded hole, to hold another part in position. If we fit a nut to a cap screw it is then called a bolt.

Cap screws are used to hold the head to the block, the crankcase cover to the crankcase, the con rod cap to the con rod. Most people refer to cap screws as bolts, so in this book we shall do the same.

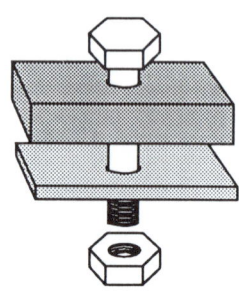
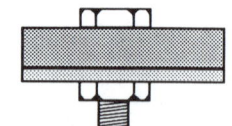

Bolt and nut

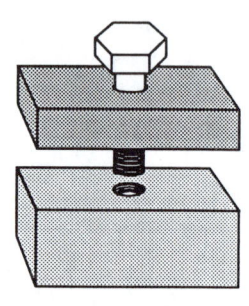

Cap screw

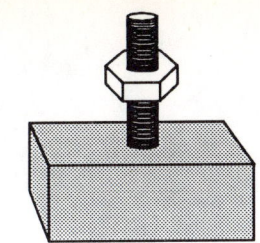

Stud

Stud

A stud is a rod threaded on both ends. Normally there is a coarse thread on the end that screws into a part and a fine thread on the other end for a nut. The stud is screwed tight into a part, then another part fitted and held by a nut. Studs are sometimes used in place of a bolt.

Washers

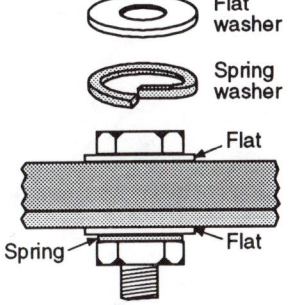

Washers

Washers are used with fasteners to improve their holding ability. Flat washers are used to distribute the clamping force and protect the parent metal.

Spring and lock type washers are used to prevent a fastener becoming loose once it is secured.

Instead of using flat washers under the head of some bolts, such as bolts to hold the head to the block, engine manufacturers are using a special bolt with a built in flange.

Keys

Flanged bolt

Keys are used to locate a part such as a flywheel, pulley or gear and in some cases to assist in transmitting power. Keys are used to locate flywheels for the timing of the spark, but the drive for the flywheel is usually through the tapered fit onto the crankshaft.

The key shape is usually square or half round (Woodruff).

Key material can be soft or hard. Always fit what the manufacturer recommends.

Threads

As mentioned in the section on hand tools there are three main threads used; British, American and metric. There are also a few manufacturers using their own special threads.

There are several different threads used by the British, American and metric systems, but we are only interested in the main ones on a small petrol engine.

Threads on bolts are generally classified as coarse or fine.

If a bolt screws into a soft metal like aluminium alloy or cast iron, it is usually coarse so that the bolt has a good anchor. If a fine thread is used it may strip the softer matching thread.

The general rule for a bolt's depth of thread into a threaded hole is that the length of the thread of the bolt into the hole should be at lease equal to twice the diameter of the bolt.

British and American Threads

In the past the British used British Standard Whitworth (B.S.W.) and British Standard Fine (B.S.F.) as their coarse and fine threads. The Americans used American National Coarse (A.N.C.) and American National Fine (A.N.F.) as their coarse and fine threads.

Later the British and Americans standardised with Unified National Coarse (U.N.C.) and Unified National Fine (U.N.F) as their coarse and fine threads which were basically of the A.N.C. and A.N.F. type. Later still the British went metric.

As a result, when a bolt disappears out of an American small engine it is not uncommon for a person to find an old Whitworth bolt in the workshop that will screw into the empty hole. However the bolt that came out was high tensile steel and the one that goes back in is probably mild steel.

Metric Threads

Metric threads originated in Europe and are spreading across the world. They are referred to as S.I. as a standard thread system. Metric threads are now standard on British equipment. The metric system has coarse and fine threads but unlike the U.N.C. and the U.N.F. where you have only one coarse and one fine thread for a given diameter bolt, the metric system can have several coarse and fine threads for a given diameter, which can cause difficulties if you are in a hurry to replace a missing bolt.

To add to the confusion there are metric bolts now available to take A.F. spanners.

Tightness of Fasteners

Threaded fasteners used on small petrol engines should be made from high tensile steel.

If you buy an ordinary bolt from a hardware store it is most probably made from mild steel. If you fit that bolt to an engine it is most likely to come loose or break.

If you want high tensile bolts you can buy them from machinery dealers or bolt specialists and they usually have markings on the head of the bolt to indicate their tensile strength.

They may also indicate what thread. U.N.C. tells you the thread, but metric symbols do not tell you the pitch.

The higher the tensile rating the higher the strength of the bolt.

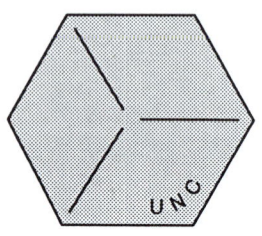

SOCIETY OF AUTOMOTIV ENGINEERS GRADE 5

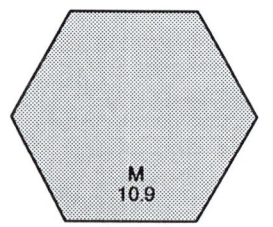

METRIC EQUIVALEN

Bolt head markings

Identifying Threads

A   Nominal diameter
B   Root diameter
C   Pitch
D   Root
E   Crest
F   Pitch angle

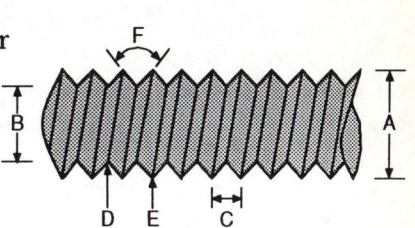

Dimensions of a Thread

Threads are accurately identified by using:

- a thread gauge
- taps and dies
- a known thread on bolts held against the unknown thread
- markings on the head of the bolt such as M for metric or UNC or UNF
- spanner fit size
- reference charts.

If replacing a nut or bolt make sure it can screw freely by hand and is not wobbly.

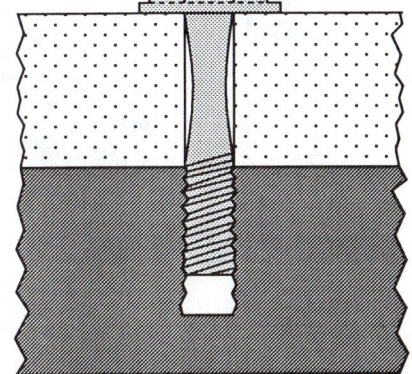

Correctly tightened bolt

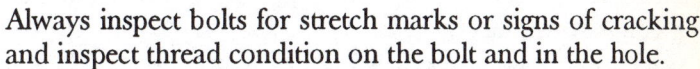

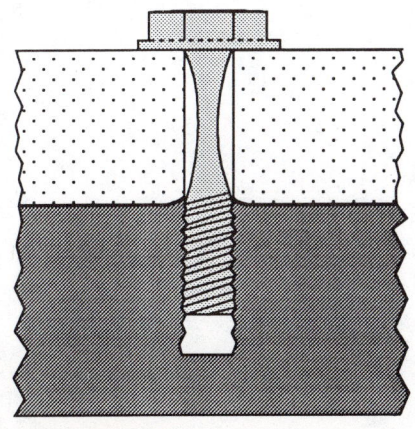

Overtightened bolt

A mild steel bolt normally has no tensile marks on its head. There is possibly a bolt manufacturer's mark on it.

Engine manufacturers have their high tensile bolts made in bulk, so they do not normally have tensile marks placed on the heads of the bolts because it is an added expense and they know where the bolt is to be used and how tight it is to go. When a high tensile bolt is tightened correctly it stretches so that tension is kept on the threads and the bolt stays tight and holds all parts in place.

Always inspect bolts for stretch marks or signs of cracking, and inspect thread condition on the bolt and in the hole.

Always ensure the right bolts go into the right holes. If a bolt needs replacing in an engine always fit the genuine article.

If a high tensile bolt is not tightened correctly, parts can move and wear. Gaskets can leak; a head gasket can blow out. The threaded hole can also wear.

If a high tensile bolt is overtightened the bolt can break or strip. The threaded hole can distort or strip out. Gaskets can leak and parts can be damaged. If it is a head bolt, the cylinder wall could distort causing piston damage, and the head gasket could blow.

Chapter 4

Engine Principles and Construction

Petrol engines are classified by a number of features.

1. Displacement, or the combined volume of cylinders:
   • measured in cubic centimetres (cc) or cubic inches (ci)
   • takes into account the bore and stroke.

2. Number of cylinders, their make and their arrangement.
   • Small engines usually have only one cylinder, larger car engines have four, six or eight.
   • A single cylinder can be positioned vertically, horizontally or at an angle.
   • Pairs of cylinders are positioned in line, in a V (usually 90°) or opposed.
   • Cylinders can be made of cast iron, aluminium alloy or a metal with a coating (e.g. chrome coating).

3. Crankshaft position can be either horizontal or vertical.

4. Cooling system type: small engines are mainly air cooled by fan action and fins; larger engines use coolants.

5. Valve location:
   • side valve (L head)
   • overhead valve (I head).

6. Camshaft location:
   • usually in the cylinder block
   • in some cases in the head (an "overhead camshaft").

7. Type of ignition:
   • usually magneto
   • points or electronically activated.

8. Number of strokes per cycle:
   • four stroke
   • some two stroke.

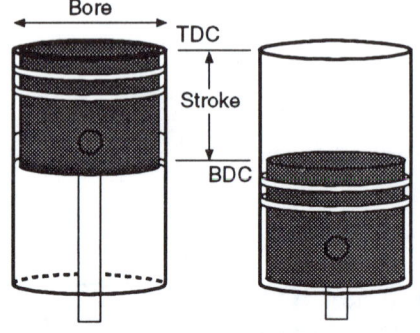

Cylinder dimensions

9. Type of fuel used:
   - petrol
   - some small engines use diesel.

10. Power rating at a set rate of revolutions per minute (rpm):
    - in kilowatts (kW)
    - in horsepower (hp).

11. Method of starting:
    - manual
    - recoil (rope)
    - electric.

Using these classifications, a small petrol engine may be described for example as a cast iron, single cylinder, four stroke, horizontal shaft, air cooled, side valve petrol engine with electronic ignition, recoil rope starting, rated at 7·5 kW at 3500 rpm with a displacement of 300 cc.

Principles of Operation

An engine can be seen as a machine where a fuel and air mixture goes in to produce the rotary motion of a shaft that we can do work with.

To assist in the understanding of the principles of operation of the engine we need to look at the elements that are required to make it work and other associated features.

The elements and associated features we will look at are:
1. Basic parts
2. Reciprocating and rotary motion
3. Air
4. Fuel
5. Combustion
6. Four stroke cycle
7. Bore and stroke
8. Displacement
9. Torque
10. Power
11. Two stroke cycle

1. Basic parts of the Engine

As was mentioned in the introduction, the early engine developers looked at the mechanics of the piston steam engine to help in the development of the internal combustion engine. The piston steam engine and the internal combustion engine use the same four basic parts to rotate a shaft to do work. These are the cylinder, piston, connecting rod and the crankshaft.

The Cylinder

- The cylinder is a round hole with parallel sides in which the piston slides up and down.
- The hole at the top is sealed off by the head and the bottom of the hole is open.

The Piston

- The piston moves up and down inside the cylinder and is a neat fit.
- The top of the piston is solid and provides the surface area on which the pressure of the expanding combustion gases pushes to force it down the cylinder.
- The underside of the piston is open.
- The piston has a hole in it so that it can be attached to the connecting rod.
- The piston has to be long enough to stop it rocking in the cylinder.

The Connecting Rod or Con Rod

- The connecting rod connects the piston to the crankshaft.
- The small end hole is attached to the piston using a pin and the big end hole is attached to the big end journal of the crankshaft.

The Crankshaft

- The crankshaft is a rotating shaft with an offset section, the big end journal, that attaches to the con rod so that it cranks when the shaft is turned.

2. Reciprocating and Rotary Motion

- The piston has reciprocating motion (up and down).
- The crankshaft has a rotary motion.

3. Air

Air is a mixture of about 23.2 per cent oxygen, 75.5 per cent nitrogen and 1.3 per cent other gases that surrounds the earth

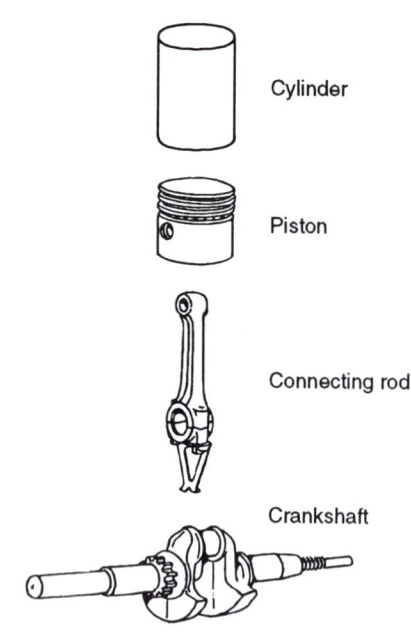

Cylinder

Piston

Connecting rod

Crankshaft

Basic parts of an engine

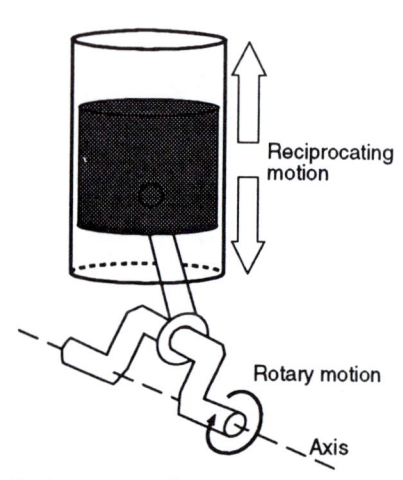

Reciprocating motion

Rotary motion

Axis

Reciprocating and rotary motion

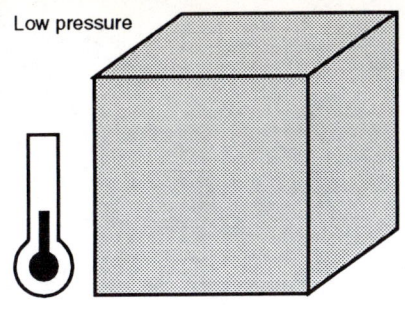

Low pressure

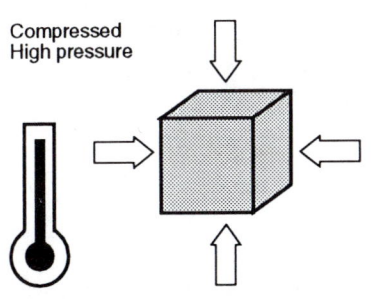

Compressed
High pressure

Air at low pressure occupies less space and has a low temperature. Air under high pressure fills a smaller space and has a higher temperature.

and forms its atmosphere. Air provides the oxygen which is needed for the burning of the fuel.

All the air surrounding the earth has sufficient weight to cause a pressure of approximately 101 kilopascals (kPa) at sea level. This is referred to as atmospheric pressure.

Atmospheric air pressure is what forces the air into the cylinder of the engine on the intake stroke.

Air will compress. Air will heat up when compressed. Air compressed into a small space has high pressure. Little or no air in the same space would be low pressure. Air will act to equalise the pressure difference if possible, flowing from areas of high pressure to areas of less pressure.

As the piston in the engine goes down the cylinder on the intake stroke it creates low pressure. With the intake valve open, the mixture of petrol and air is pushed into the cylinder by the outside atmospheric air pressure.

4. Fuel

The most commonly used fuel is petrol. Petrol is the trade name of a fuel that is obtained by refining crude oil to obtain specific hydrocarbons (compounds of carbon and hydrogen atoms). Some specific additives are also included, producing two common grades, super and unleaded. Super is gradually being phased out.

Petrol contains chemical energy which is released when it is burned. It is a volatile fuel that mixes easily with air (when it vaporises) and ignites easily.

For the petrol to burn inside the engine's cylinder it must be mixed finely with air at a certain ratio. This ratio is given by weight and is approximately 15 parts of air to one of petrol for normal running. If it were given as a volume it would be approximately 9000:1.

The petrol is atomised in the carburettor before it enters the cylinder. The petrol and air mixture that leaves the carburettor is in the form of a fine spray, similar to the spray that comes out of a pressure pack fly spray.

Before this mixture is burned in the cylinder it must be a vapour so that the maximum amount of energy is obtained from the petrol. Vaporisation of the petrol starts as it mixes with the air flow in the carburettor and is assisted by the heat of the intake system and by the heat from the compression stroke in the cylinder.

5. Combustion

Combustion is a constant burning action that occurs when a mixture of fuel and air is ignited.

When combustion of a volatile fuel like petrol occurs in a confined space like a cylinder, pressure is produced to drive the piston down.

The engine developers found, by a process of trial and error, that three things played a major role in the combustion process:
1. The compression of the air.
2. How volatile the fuel used was.
3. The amount of fuel used.

Engine developers such as Otto found that if they compressed the petrol/air mixture prior to ignition they could raise the pressure of the mixture, which also increased its temperature. When the mixture was ignited the resulting combustion would produce higher temperatures to give higher pressures to push the piston down with greater force, and more power could be obtained from the engine.

If air was compressed too much in the engine cylinder, the heat of the air would ignite a fuel if it was placed in it. This is the principle of the compression ignition (diesel) engine, where air is compressed to a ratio of about 16:1, then diesel fuel is injected into the cylinder.

Because petrol is mixed with the air prior to entering the petrol engine cylinder, the petrol and air are compressed together. If it were compressed to a 16:1 compression ratio like the diesel engine the mixture would self ignite before the piston reached the top of the cylinder. As a result uncontrolled combustion would take place, the engine would not be easily controlled and damage would result.

Engine developers found they could safely compress the petrol and air to a ratio of about 9:1. Some performance engines run higher ratios, but generally a ratio of about 8:1 or 9:1 is used for overhead valve engines and about 6:1 for side valve engines.

Compression ratios tell us how much the mixture is compressed by volume. It is the volume in the cylinder when the piston is at the bottom of its stroke compared to the volume that is left in the cylinder when the piston is at the top of its stroke.

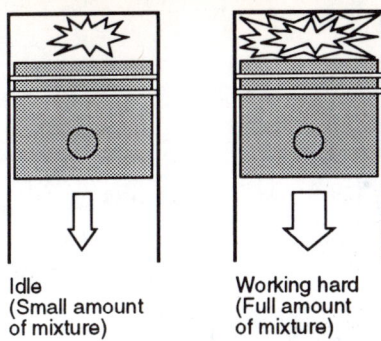

Idle
(Small amount
of mixture)

Working hard
(Full amount
of mixture)

Idle and full throttle mixtures

Engine performance depends on the two factors:

- How volatile the fuel is: a volatile fuel is one that turns to vapour easily. As a vapour, petrol will ignite more readily and burn more quickly. In the carburettor, atomising petrol in air and compressing it help make the fuel efficient.

- The amount of fuel used: to control combustion, which in turn controls the power output of the engine, you must be able to control the amount of mixture that goes into the cylinder.

At idle, only a small amount of mixture is needed to keep the engine running.

When the engine is working hard you need the maximum amount of mixture you can fit into the cylinder, so that combustion produces maximum pressure to push the piston down with maximum force to rotate the crankshaft at a given speed to produce the required power.

The amount of mixture entering the engine is usually controlled by the throttle butterfly valve in the carburettor.

6. Four Stroke Cycle

This is the series of events that places a fresh mixture of petrol and air in the cylinder, compresses it, ignites it, burns it, forces the piston down to do work, removes the burnt gases, then repeats the cycle so as to keep the crankshaft rotating.

The principles of Dr Otto's four stroke cycle engine of 1876 are still in practice today. The two stroke cycle completes the series of events in two strokes of the piston. The four stroke engine is the most common and therefore we will mainly concentrate on it.

A stroke of the piston refers to its movement as it travels from one end of the cylinder to the other.

Each stroke requires the crankshaft to rotate half a turn (180°). During the four stroke cycle the crankshaft rotates twice (720°). There are two up strokes and two down strokes in the cycle.

The four strokes in sequence are:
- intake (down)
- compression (up)
- power (down)
- exhaust (up).

The four stroke engine

Head: to seal off the top of the cylinder to provide a combustion chamber.

Piston rings: to provide better sealing between the piston and cylinder.

Spark plug: to provide a spark to ignite the mixture.

Inlet valve: through which the fuel mixture is admitted to cylinder.

Inlet port: to provide a passage for the fuel mixture to enter the engine.

Exhaust valve: through which burnt gases pass out of the cylinder.

Exhaust port: to provide a passage for burnt gases to leave the engine.

Camshaft: to provide means to open the valves at the correct time (driven by the crankshaft).

Valve springs: to close the valves.

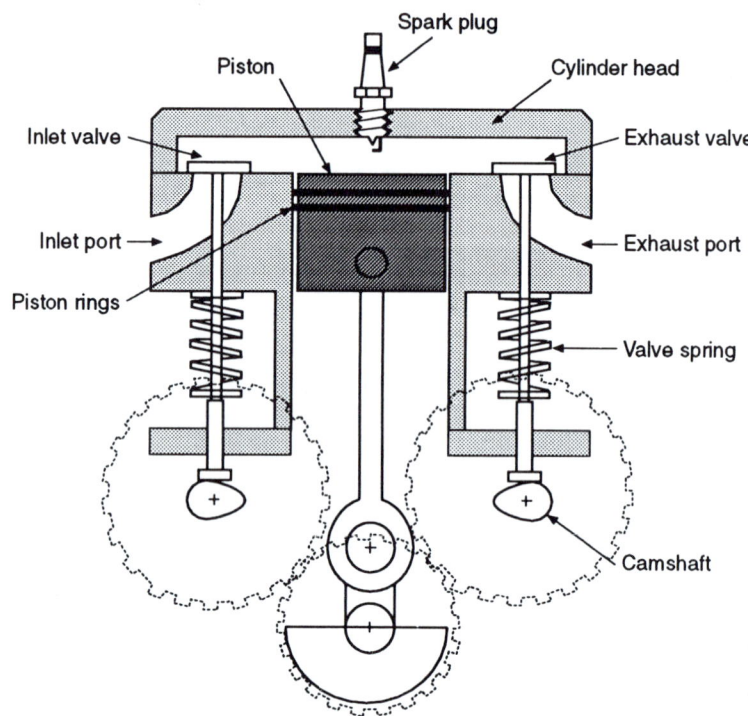

Intake Stroke

On this stroke a fresh petrol/air mixture enters the cylinder. The piston travels from the top of the cylinder to the bottom, which is half a turn of the crankshaft. The inlet valve opens while the exhaust valve is closed. As the piston descends it reduces pressure in the cylinder. This allows the outside atmospheric air pressure to push air through the carburettor, collecting petrol. The mixture enters the cylinder through the opened inlet valve.

At the end of the intake stroke the cylinder has filled up with a fresh petrol/air mixture. Shortly after, the inlet valve closes trapping the mixture in the cylinder.

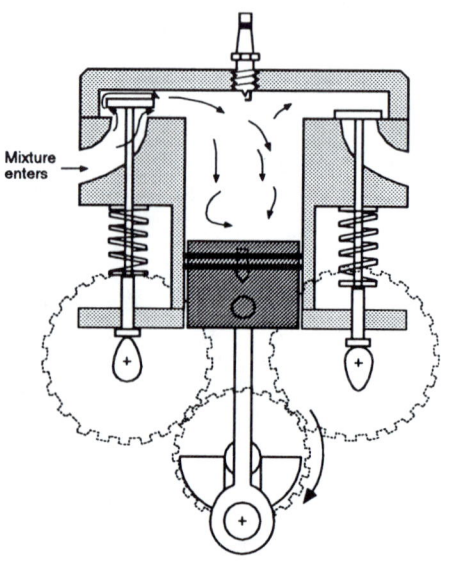

Compression Stroke

On this stroke the mixture is compressed into the space at the top called the combustion chamber. Both the inlet and exhaust valves are closed. As the piston travels from the bottom of its stroke to the top, the mixture is compressed to about one sixth of its original volume. As a result the pressure and temperature rise. Near the end of the compression stroke, ignition of the mixture takes place. A spark at the spark plug starts the combustion process. The mixture then burns across the combustion chamber generating heat. The heat and combustion gases are trapped inside the combustion chamber and there is an increase in pressure. This pressure is exerted evenly in the combustion chamber, and forces the piston down by pushing on the surface area of the top of the piston.

The combustion process starts just before the very top of the stroke and finishes just after so the expanding gases can push down on the piston to produce work.

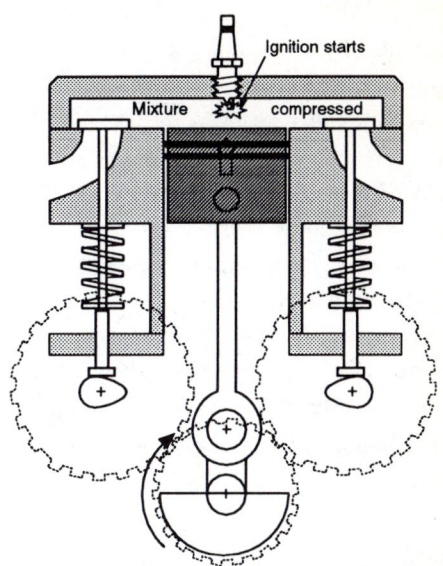

Power Stroke

On this stroke the piston is forced down the cylinder by the pressure of the expanding combustion gases. Both valves are closed so the pressure is sealed in. By the time the piston is about half way down the cylinder the maximum leverage is exerted onto the rotating crankshaft to produce work. The leverage effect is the torque or twisting effort of the crankshaft which we shall look at shortly.

The power stroke is also sometimes referred to as the expansion, firing or working stroke.

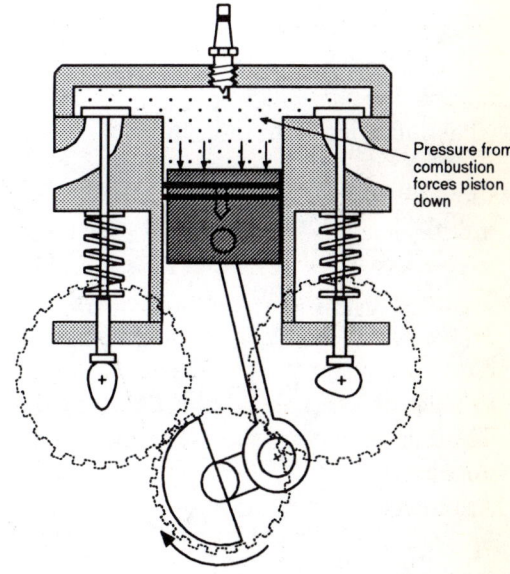

Exhaust Stroke

Near the end of the power stroke the exhaust valve starts to open so that the burnt gases can then be released to the atmosphere. The inlet valve stays closed. When the valve is opened, the internal pressure, which is higher than the atmospheric pressures, will cause the burnt gases to depart.

The rising piston pushes the burnt gases out of the cylinder as it travels up the cylinder. All the exhaust gases cannot be expelled on a small petrol engine, so a portion is left in the combustion chamber when the piston is at the top.

The exhaust stroke is the end of the four stroke cycle. The next stroke is the inlet stroke and the cycle repeats itself.

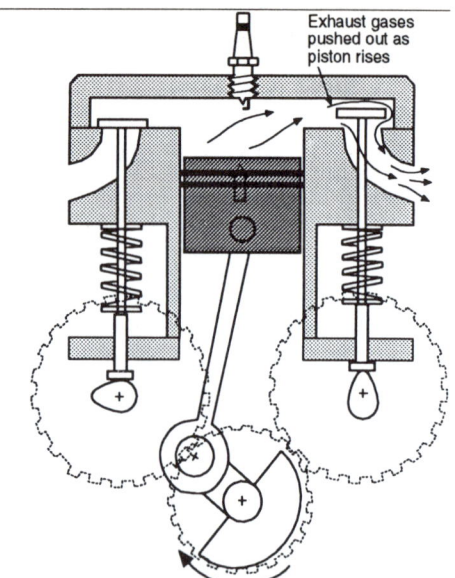

Exhaust gases pushed out as piston rises

Valve Operation

The inlet and exhaust valves do not open and close exactly at the start and finish of their respective stroke, but open and close near the start and finish. The engine manufacturers experiment with the valves so that the mixture has the maximum time to enter the cylinder and likewise the exhaust gases to leave the cylinder.

The following is an indication of the valve operation:

Stroke	*Inlet valve*	*Exhaust valve*
Intake	Open	Closes at start
Compression	Closes at start	Closed
Power	Closed	Starts to open near end
Exhaust	Starts to open near end	Open

The valves are operated by the camshaft which is timed to and driven by the crankshaft. The camshaft is geared to rotate at half the speed of the crankshaft. It is driven either by a gear or chain drive.

7. Bore and Stroke

The bore is the diameter of the engine cylinder. It is expressed in millimetres.

The stroke is the distance the piston moves from top dead centre (T.D.C.) to bottom dead centre (B.D.C.). It is expressed in millimetres.

Offset on the crankshaft is half the stroke.

8. Displacement

This is also called engine capacity or swept volume.

In the case of a small one cylinder engine, the displacement is the total working volume of the cylinder.

The displacement is the difference between volume when the piston is at the top and when it is at bottom of its movement.

The volume of any cylinder equals $3.142 \times radius^2 \times height$. In an engine cylinder this is

$$3.142 \times \left(\frac{bore}{2}\right)^2 \times stroke$$

The displacement of a cylinder with a bore of 80 mm and a stroke of 60 mm is:

$$Displacement = 3.142 \times \left(\frac{80}{2}\right)^2 \times 60 \text{ mm}$$

$$= 301632 \text{ mm}^3$$

$$= 302 \text{ cubic centimetres (cc)}$$

If you took the head off this engine and turned it from T.D.C. to B.D.C. you would be able to pour in 302 cc of oil into the cylinder to indicate displacement. This exercise is handy if you are trying to identify the displacement of an engine.

For multiple cylinder engines, multiply displacement of one cylinder by the number of cylinders to obtain engine displacement.

9. Torque

Torque is a twisting or turning effort. It is measured in Newton metres.

A Newton is a unit of force. If 1 Newton of force is applied to the end of a lever 1 metre from an attached pivot the torque effort at the pivot would be 1 Newton metre.

In the engine, combustion provides the pressure to push on the piston to force it down the cylinder. On the diagram this force (A) is then applied to the offset big end journal of the crankshaft (B) through the connecting rod (C). The force (A) acts over the distance that the centre of the offset big end journal is from the centre of the crankshaft (D). This acts as a lever to rotate the crankshaft. Maximum lever action (torque) is attained when the connecting rod is at 90° to the centre line of the big end and main journals of the crankshaft, when the piston is about half way down on the power stroke.

Engine manufacturers usually have a chart in the form of a graph to show the torque and power of their engines so that people can apply their engines to a particular purpose. They may use two separate graphs, as shown, or they may combine them.

Maximum engine torque in a small petrol engine is usually reached at about 2800 rpm with a maximum working rpm of between 3500 and 4000 rpm.

Either side of the 2800 rpm mark the engine does not breathe as well and torque drops off. Small petrol engines usually run at revolutions higher than the maximum torque figure so that when extra load is applied the engine slows down slightly and the torque increases and the engine "holds on".

Torque is related to power, which is a more useful way of rating an engine.

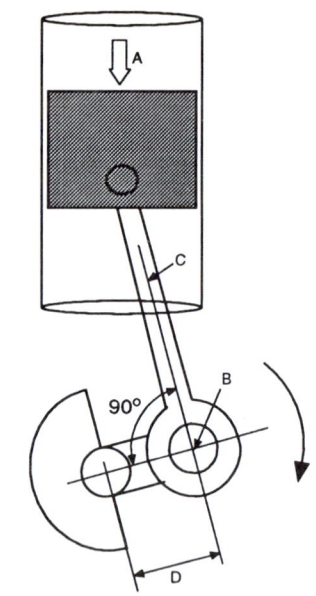

Maximum torque is applied to the crankshaft about halfway down on the power stroke

10. Power

Power is the rate at which work is done. When people go to buy an engine they are usually concerned with the horse-power of the engine at certain rpm. Power is measured in watts; the most commonly used unit is kilowatts, but many people still use horsepower. One horsepower is the energy required to lift 33 000 pounds through a distance of one foot in one minute.

The power output of an engine is related to the torque of the engine at speed. As was noted in the torque graph the torque peaks at about 2800 rpm then drops off.

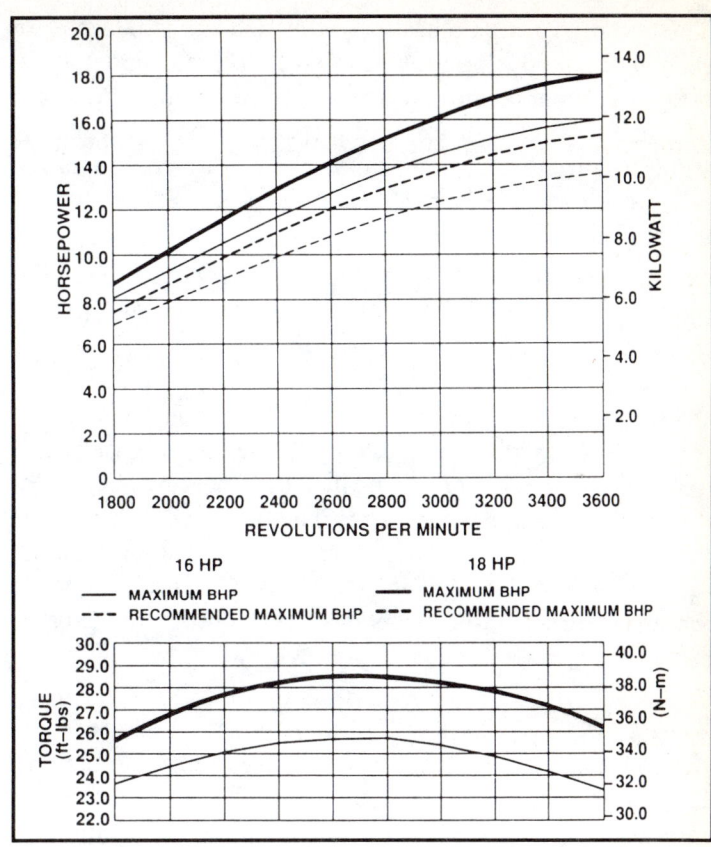

An example of an engine manufacturer's torque and power graph

On the power graph the power climbs, then peaks at about 4000 rpm, then starts to drop. This is related to the drop in torque at higher engine speeds.

The graph also shows a line for the maximum Brake Horsepower (B.H.P.) and the recommended B.H.P. The maximum B.H.P. is what the engine can produce on the dynamometer but it is recommended to use the lower B.H.P. figure in general use, mainly to make the engine work more easily and last longer. The engine is probably advertised at the higher rating.

The suitability of engines of varying power is discussed in Chapter 14.

11. Two Stroke Cycle

The two stroke engine's main use is in chainsaws and small outdoor power products. It is also used in some motorbikes and lawnmowers.

The two stroke engine has to go through the same series of events: place a fresh petrol/air mixture in the cylinder, compress it, ignite it, burn it, force the piston down to do work, then remove the burnt gases and repeat the cycle, to keep the crankshaft rotating. But a two stroke engine completes the series of events in only two strokes of the piston (one turn of the crankshaft).

The two stroke engine utilises ports and valves (reed and rotary types) to let the mixture into the engine and one or more ports to let the burnt gases out.

Ports

A port is a hole or opening in the cylinder, whose use is controlled by the movement of the piston. When the piston is at the top of its stroke, it closes off the ports, sealing and compressing the fuel gases. As it descends after ignition it passes the ports, effectively opening them to let fresh fuel mixture into the cylinder, and exhaust gases out through the exhaust port.

Reed Valves

A reed valve is similar to a one way valve used as a foot valve for pumping water. It is usually located between the crankcase and carburettor. The valve itself is a sheet of spring steel. The valve is pushed open by atmospheric pressure which allows the mixture to enter the crankcase. It is closed by the action of the spring steel and crankcase pressure.

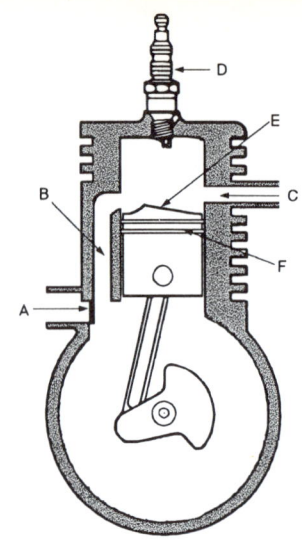

A Reed valve, to allow the fuel mixture to enter the crankcase and stop it escaping back to the carburettor. (Some engines use ports instead of a reed valve or they use both).

B Transfer port, to allow the mixture to enter the cylinder above the piston.

C Exhaust port, to allow the burnt gases to leave the cylinder to the atmosphere.

D Spark plug, to ignite the mixture for combustion.

E Contoured top of piston, to direct the mixture into the cylinder and the exhaust gases out.

F Piston rings, to give a good seal between the piston and the cylinder.

Two Stroke Operation Principles

In a two stroke engine the crankcase has an extra role. It is used to create high and low pressures to pump the mixtures from the carburettor to inside the cylinder via the crankcase.

First Stroke:

As the piston went up the cylinder on the previous stroke, the volume in the crankcase increased, creating low pressure. This allows the outside atmosphere pressure to push the petrol/air mixture through the carburettor and against the outside of the reed valve, forcing it open. This fills the crankcase with fuel mixture. At this point ignition takes place in the cylinder and the piston descends. Near the bottom of the stroke the exhaust port is uncovered, allowing the exhaust gases to start escaping.

As the piston moves down the cylinder, it decreases the volume in the crankcase which creates an area of high pressure. This high pressure assists in quickly closing the reed valve (which also is closed by its spring) near B.D.C. The pressurised mixture tries to escape, but there should be no outlet for it until the transfer port is uncovered by the piston when it nears B.D.C.

When the transfer port is uncovered, the high pressure in the crankcase pushes the mixture into the cylinder. The contoured top of the piston directs the mixture upwards, it hits the head, then loops downwards. The contour on the piston on the exhaust side directs the exhaust gases out the exhaust port as the fresh mixture enters to replace it. This is the "scavenging" effect of the incoming mixture cleaning out the burnt gases.

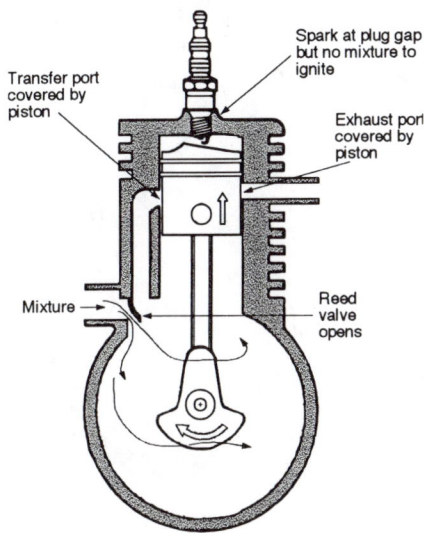

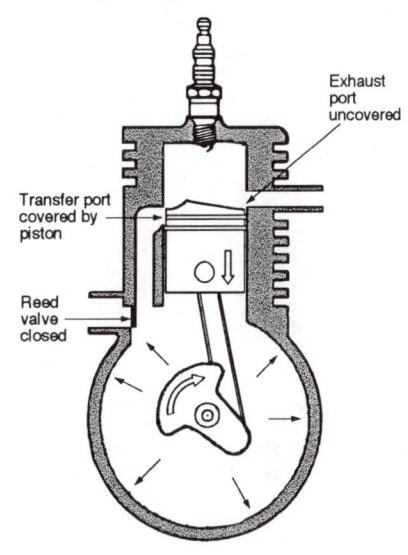

Second Stroke:

As the piston moves up the cylinder, it first covers up the transfer port, then the exhaust port. It then compresses the fuel mixture in the cylinder. Just before T.D.C. the spark at the plug gap ignites the mixture to start combustion.

Combustion finishes just after T.D.C. In the crankcase a new mixture is being pushed through the reed valve due to the low pressure that was created by the upward travel of the piston.

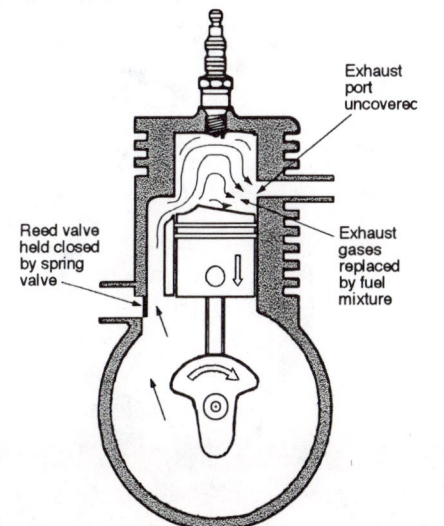

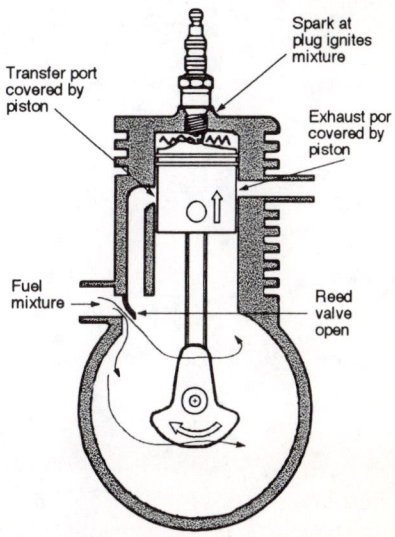

Operating cycle of a two stroke engine using ports only.

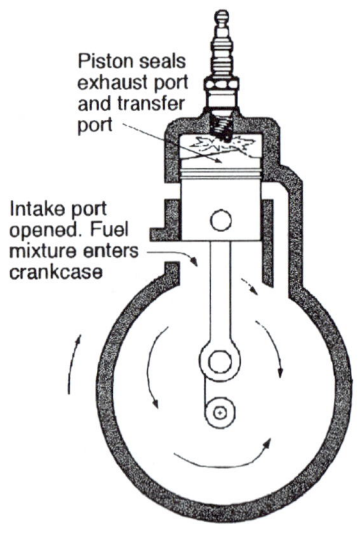

Piston seals exhaust port and transfer port

Intake port opened. Fuel mixture enters crankcase

Ignition

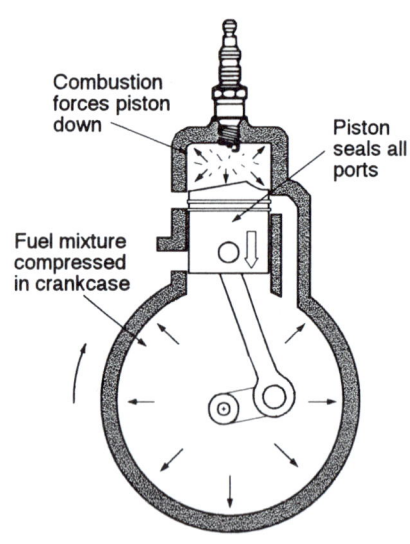

Combustion forces piston down

Piston seals all ports

Fuel mixture compressed in crankcase

Power

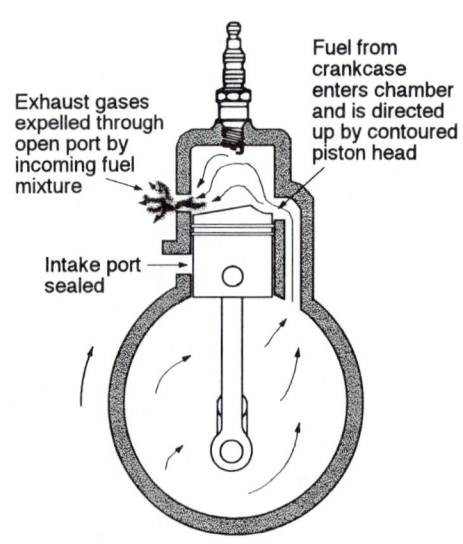

Exhaust gases expelled through open port by incoming fuel mixture

Fuel from crankcase enters chamber and is directed up by contoured piston head

Intake port sealed

Exhaust

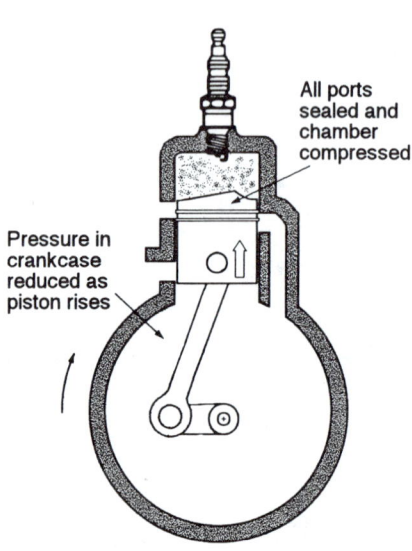

All ports sealed and chamber compressed

Pressure in crankcase reduced as piston rises

Compression

In theory a two stroke engine of the same displacement as a four stroke engine should deliver twice the power. In reality this is not the case due to a number of factors, a main one being that the "scavenging" effect of the two stroke is not very efficient.

The two stroke can be made more compact and lighter than a similarly power-rated four stroke engine and is therefore ideal for chainsaws and similar products. Two stroke engines also have a lot fewer moving parts than a four stroke.

However, the two stroke requires oil to enter the crankcase with the petrol/air mixture to provide lubrication and is not as fuel efficient as a four stroke.

Engine Construction

If a small petrol engine were pulled apart, it would look something like this:

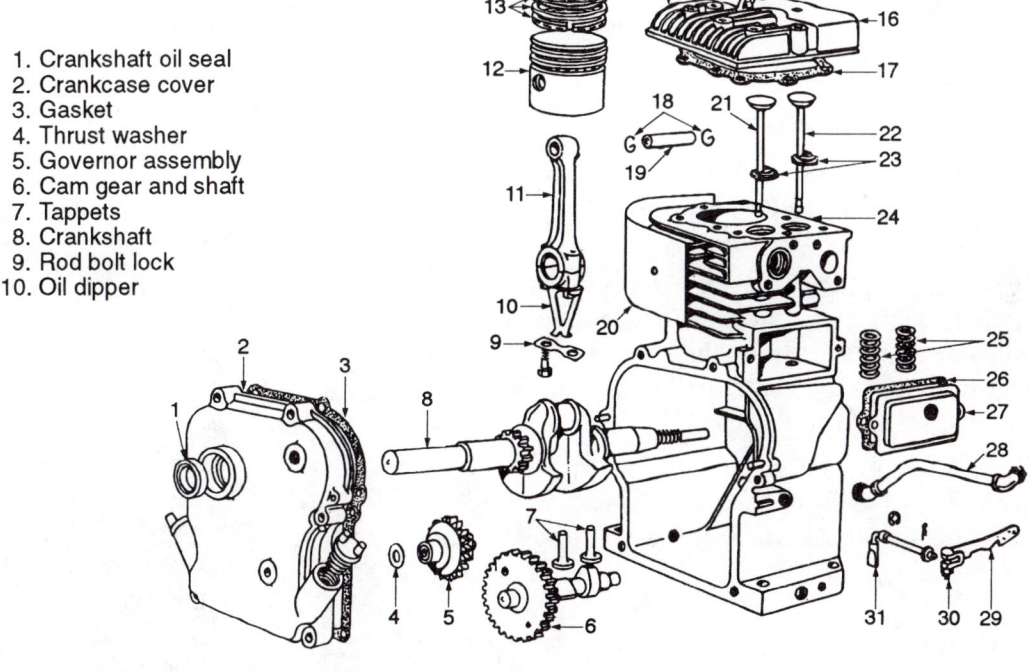

1. Crankshaft oil seal
2. Crankcase cover
3. Gasket
4. Thrust washer
5. Governor assembly
6. Cam gear and shaft
7. Tappets
8. Crankshaft
9. Rod bolt lock
10. Oil dipper

11. Connecting rod	16. Air baffle	21. Exhaust valve	27. Breather & tappet
12. Piston	17. Cylinder head gasket	22. Inlet valve	chamber cover
13. Piston rings	18. Piston pin retaining	23. Valve spring retainers	28. Breather pipe
14. Cylinder head	rings	24. Cylinder block	29. Governor lever
15. Spark plug ground	19. Piston pin	25. Valve springs	30. Clamping bolt
switch	20. Air baffle	26. Gasket	31. Governor crank

We will now look at the main parts that go to make up a four stroke, air cooled side valve small petrol engine and discuss the role of each in the mechanical system. We will then assemble the parts together to make an engine.

The Cylinder Block

This is the large main part of the engine. The top section contains the cylinder and the bottom section the crankcase. All other parts of the engine are found working in connection with it or attached to the inside or the outside of it.

The shape is complicated and is cast in a mould as a single unit. It is usually made from cast iron, aluminium alloy or other light metals. After the block is cast, certain sections are then machined; holes are drilled and threaded.

In the top section of the block are the:
• cylinder
• inlet and exhaust valve, with seats, ports, guides and tappet chamber
• machined section for attaching the head, intake system and exhaust system
• cooling fins around outside.

The Cylinder

This is the round hole with smooth parallel sides that is found in the cylinder block. The piston slides neatly up and down inside the cylinder.

The cylinder is made of the same material as the block, usually cast iron or aluminium. A cast iron cylinder sleeve can be fitted into the mould for an aluminium block and becomes an integral part of that block. Some cylinders may even have a coating of a metal such as chrome.

Before the piston is fitted into the cylinder, the cylinder is machined and honed to the required bore size for the engine. The finish to the working surface of the cylinder is a cross hatch pattern to aid piston ring seating and sealing.

The Crankcase

The crankcase is the lower part of the cylinder block in which the crankshaft is housed and revolves.

In the crankcase are also be found the
• camshaft and tappets
• timing gears

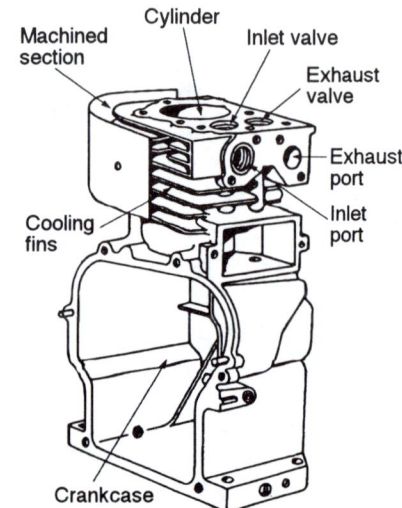

Machined section · Cylinder · Inlet valve · Exhaust valve · Exhaust port · Inlet port · Cooling fins · Crankcase

* lubrication system
* governor assembly.

On the other side of the crankcase can be found:
* mounting points for the ignition components
* a hole for the end of the crankshaft to come through to rotate the flywheel.

The Crankcase Cover

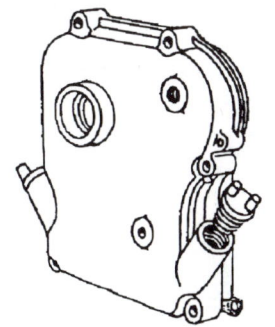

This is the cover that encloses the crankcase. It provides an end support for the camshaft and crankshaft and a hole for the crankshaft to exit the engine. It keeps the oil in and the dirt out. Some older engines had open crankcases.

The Cylinder Head

The head is a metal cover bolted to the top of the cylinder block to enclose the top of the cylinder. Most cylinder heads are designed to be removable so that work can be carried out easily on the top end of the engine.

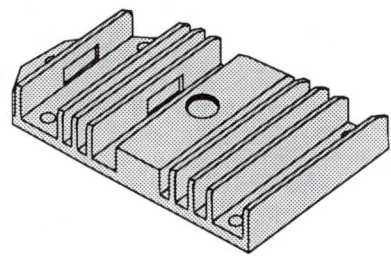

By removing the head:
* valves can be serviced
* pistons can be removed
* carbon can be removed.

The head is usually cast from either aluminium alloy or iron.

Features of the head:
* a threaded hole for the spark plug
* holes for bolts to go through to secure it to the block
* cooling fins
* hollowed underneath to provide a combustion chamber and room for valves to open
* flat section to be bolted to the top of the block for a good seal.

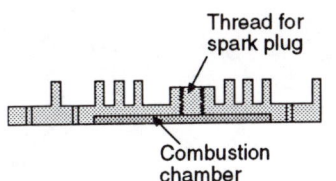

Thread for spark plug

Combustion chamber

The Cylinder Head Gasket

The head gasket is a gasket that is placed between the top of the cylinder block and the base of the cylinder head to ensure good sealing. It is made from material that will withstand the heat and pressure of combustion and must be tightened evenly to ensure a good seal.

The Connecting Rod

The connecting rod is the connecting link between the piston and the crankshaft. It is often referred to as the con rod.

The con rod transfers the force from the piston to the crank-shaft. It is usually either cast aluminium alloy or forged steel.

The top of the rod that attaches to the piston is referred to as the small end.

The bottom of the rod that attaches to the offset journal of the crankshaft is usually referred to as the big end of the rod and contains the bearing surface to run on the big end journal of the crankshaft.

The bottom of the rod has a removeable cap so that it can be easily fitted to the crankshaft. This cap is matted to the rod and only fits on one way. It usually has identification marks to ensure correct fitting.

The Piston

The piston is a cylindrical piece of metal that fits neatly into the cylinder and moves up and down.

Its main job is to provide a surface area for the pressure of combustion to push against so the resultant force applied to the piston can be transferred to the connecting rod.

The piston is cast or forged and usually made from aluminium alloys for lightness and efficiency. Older slow working engines used cast iron pistons which work well, but when engines were speeded up the heavy pistons wore out the bearings.

The hole for the piston pin may be offset slightly to one side of centre to reduce the 'slapping' effect and the wear characteristics in the cylinder. The piston must go into the cylinder the correct way around.

To provide for better sealing and oil control between the piston and cylinder there are piston rings fitted, usually two compression rings and one oil ring. Grooves are machined into the piston to accept the rings.

To stop the piston pin from moving sideways and damaging the cylinder wall the piston has grooves machined to take retaining clips. The piston is not machined to a round finish. If it were, it would change to an egg shape when it got hot, due to the extra metal around the piston pin boss area expanding, and the difference in temperature between the top and bottom of the piston. The manufacturers machine the piston to a certain oval shape (called "cam ground") when viewed from the top and a slightly tapered shape when viewed from the side. When it reaches operating temperature it fits neatly into the cylinder.

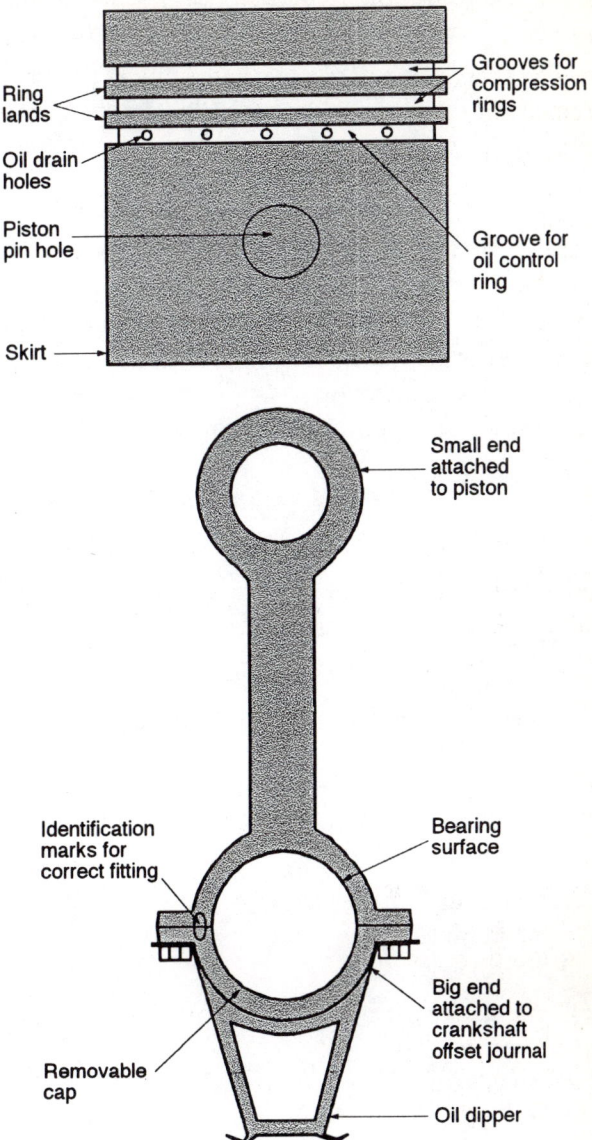

Ring lands
Grooves for compression rings
Oil drain holes
Piston pin hole
Groove for oil control ring
Skirt
Small end attached to piston
Identification marks for correct fitting
Bearing surface
Big end attached to crankshaft offset journal
Removable cap
Oil dipper

Pistons may also have steel inserts and slots to control expansion. Some pistons, such as those used in aluminium alloy bores have a special coating such as tin.

The Piston Pin

The piston pin is a pin that connects piston and con rod.

It is a very neat fit and most small petrol engines use the type that is called fully floating; a neat sliding fit in both the piston and the con rod.

The pin has to be able to handle heavy loads so it is made from good quality steel, case hardened, hollow, and highly polished. Circlips are used to stop it drifting out of the piston where it could badly damage the cylinder wall.

The Piston Rings

Because the piston itself can not seal well enough in the cylinder, piston rings have to be used.

The rings are designed to:
* be squeezed into the cylinder so that pressure from the ring is exerted against the cylinder wall at all times as the piston goes up and down
* fit neatly into the groove in the piston with a minimum of side clearance
* always have back clearance to allow for the expansionof the variuos components.

The rings are circular so as to provide a long lasting type of seal and have a gap so as to allow for expansion when hot. The gap will close slightly when hot but should never meet, otherwise ring breakage could occur.

Piston rings are usually made from cast iron. The rings can therefore be easily broken whilst expanding them to slip them onto the piston if care is not taken. Rings may also have a surface coating such as chrome. Chrome rings should never be used in chrome cylinders.

There are two main types of rings:
* compression rings
* oil control rings.

Small four stroke petrol engines usually have:
* two compression rings
* one oil control ring.

Compression Rings

Their job is to reduce leakage between the piston and the cylinder wall by providing a good seal. Various shapes are used such as shown in the illustration.

The rings are usually marked to identify which way is up for correct installation.

On the power stroke the high pressure of the combustion goes down the side of the piston and pushes the compression rings down against the piston and out against the cylinder wall for a good seal, so that the combustion pressure can

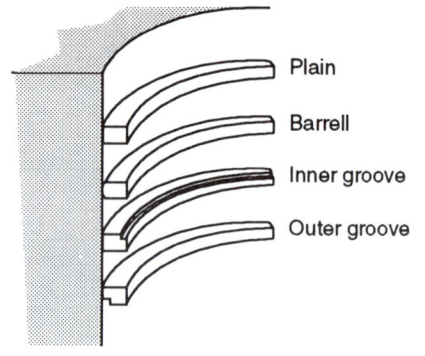

Different types of compression rings

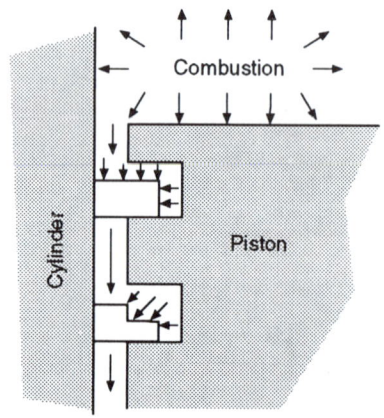

Compression rings form a seal between piston and cylinder wall

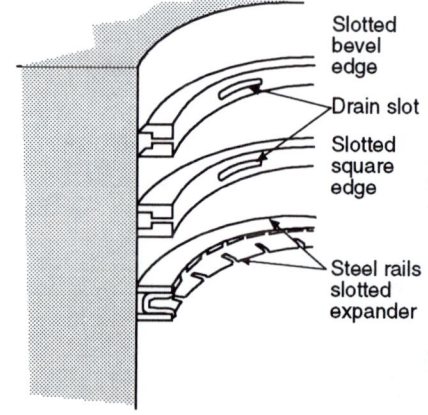

Different types of oil rings

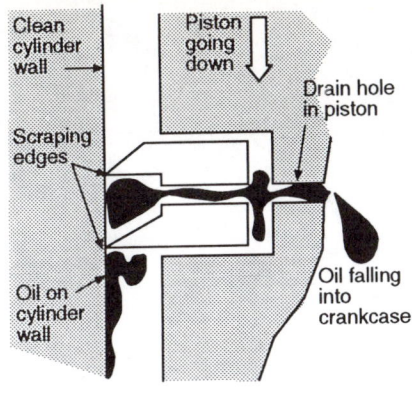

have maximum effect on the surface area of the top of the piston, to force it down.

The pressure that leaks past the rings is known as 'blow by'.

Oil Control Rings

The oil control ring is used to scrape the surplus oil from the cylinder wall.

There are various types, with the types used on small petrol engines usually being of one piece construction. Some rings have spring expanders behind them to help exert more pressure against the cylinder wall.

The oil control ring usually rides over the oil as it heads upwards and when it travels down it scrapes oil off, draining through holes in the piston or back down the cylinder wall.

A certain amount of oil has to lubricate the cylinder wall or damage could result.

The Crankshaft

This is the main shaft in the engine, that changes the reciprocating motion of the piston into rotary motion to provide a constant turning action. The reciprocating motion of the piston is transferred to the offset pin on the crankshaft by the connecting rod.

The crankshaft is located in the crankcase and the ends of the shaft protrude out of the crankcase. It is supported by the main bearings on either side of the offset crankpin. The bearings may be:
*   plain bearings machined as part of the parent metal
*   removeable type like a bush
*   anti-friction type like a ball or roller bearing.

The main bearings also control the end float of the crankshaft, and have provision for lubrication from the oil in the crankshaft and room to fit an oil seal to the outside area.

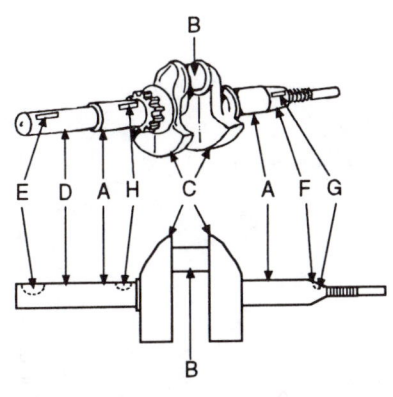

A Main journals
B Big end journal
C Counterweights
D Output drive end
E Keyway for drive end
F Flywheel end
G Flywheel keyway
H Keyway for timing gear (some crankshafts have the timing gear cast as part of the crankshaft)

The crankshaft is usually of one piece construction and made from either:
*   cast iron, or
*   forged steel.

The journals, pins and throws are precision machined.

The offset section to which the connecting rod is attached is often referred to as the big end journal, the crank pin, the crank journal, the rod journal or the con rod throw.

If we use a simple crankshaft design the engine would vibrate badly when running because the offset section puts the shaft out of balance. To correct this, counterweights are situated opposite the offset section. The flywheel also helps to smooth out some of the vibration. Some engines also use balance shafts or other balancing devices. The more cylinders an engine has, the lower the vibration level.

The Camshaft

The camshaft is the shaft in the engine which pushes open the intake and exhaust valves.

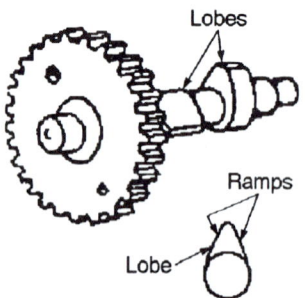

The camshaft is usually made from cast iron. It may have an integral timing gear or one which is removable.

The camshaft has two lobes or cams. The lobes provide a ramp to push against the tappets or valve lifter, to cause the valve to open.

When the camshaft is made the lobes are machined so that the intake and exhaust valve open correctly and the right distance.

Valve springs pull the valves closed.

Valve Timing

The intake and exhaust valves are timed to the movement of the crankshaft and piston. The camshaft is driven by the crankshaft to ensure the valves open at the correct time. The camshaft rotates at half engine speed.

Gears or sprockets and chain provide the means of drive between the crankshaft and camshaft. Gears are the most common method used. The gear on the crankshaft is half the diameter of the gear on the camshaft and has half as many teeth. The crankshaft gear and the camshaft gear are timed to each other usually with visible timing marks.

If we rotate the crankshaft one full turn the crankshaft timing gear turns 360°, the camshaft timing gear turns 180° and the timing marks are opposite each other. If we rotate the crankshaft one more full turn the crankshaft timing gear turns another 360°, the camshaft timing gear turns another 180° and the timing marks line up again.

The Valve System

The engine valve is a device used to control the flow of gases through a hole.

The engine uses two valves:
- The intake valve controls the flow of mixture entering the cylinder.
- The exhaust valve controls the flow of burnt gases leaving the cylinder.

The valves are opened by the camshaft lobes and closed by the valve springs.

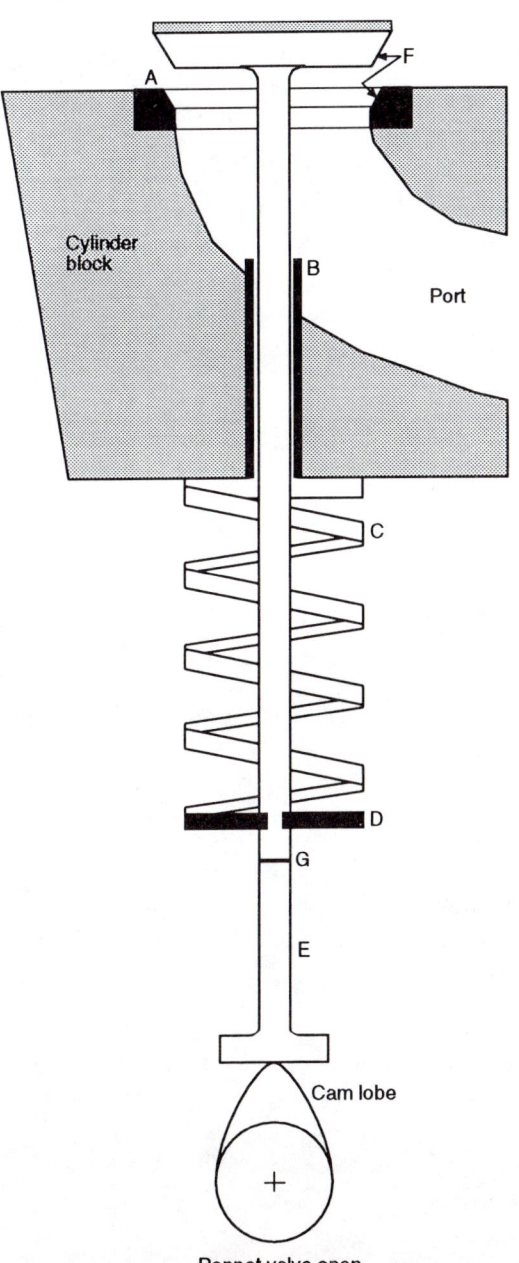

Poppet valve open

A Valve seat insert. A hard steel circular ring that is inserted mainly into aluminium alloy cylinder blocks. It provides a seat for the valve face to seal upon. Cylinder blocks made from cast iron may use the parent metal as the seat.

B Valve guide. A bush that is pressed into most aluminium cylinder blocks, so that the stem of the valve can slide up and down in it and be guided accurately, so that it seals well. Cast iron cylinder blocks may use the parent metal as a guide. Heat from the valve head also travels down the stem and exits at the valve guide.

C Valve spring. A strong spring which closes the valve once the cam lobe allows the tappet to run down the closing ramp.

D Valve spring retainer. A device secured to the end of the valve stem to hold the valve spring in place. Some engines use rotators on their exhaust valves.

E Tappet. Also called valve lifter or cam follower. This device is lifted as the cam rotates and opens the valve. It has a large base so that the cam lobe bears evenly against it to reduce wear.

F The area where the valve face contacts the valve seat to provide the sealing effect. Another job of this contact area is to allow heat from a valve, especially an exhaust valve, to be conducted onto the cooler insert and block.

G Tappet clearance. When the valve is fully closed, there must be a clearance in the valve mechanism and it is usually checked here. If there is no clearance the valve face will not correctly contact the seat and the valve can burn out, resulting in loss of compression. If there is too much clearance the valve timing will be out and the engine will not run very well. Adjustment is normally done by grinding the end of the valve stem. Some engines have threaded adjusters.

The commonly used valve is of the poppet design. This design has been around over 100 years. Other designs have been tried but the poppet valve is still favoured. It is so named because it pops up and down as it works. It is also called a mushroom valve.

Valves are made from good quality steel. The exhaust valve must be made extra well because it operates at high temperatures when the engine is working hard. The inlet valve operates in reasonably cool conditions compared to the exhaust valve because it has the incoming mixture passing over it all the time.

When closed the valve face seals against the valve seat in the cylinder block.

The valve stem moves up and down in the valve guide.

The valves are free to rotate as they work which helps clean the face for a good seal.

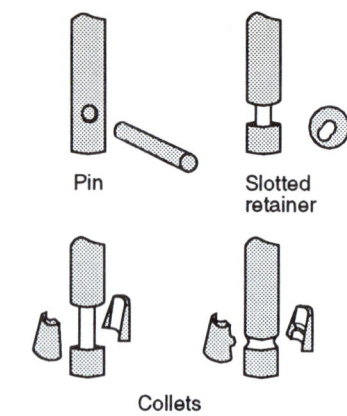

Pin   Slotted retainer

Collets

Methods used to retain valve springs

The Flywheel

The flywheel is a heavy wheel that is attached to one end of the crankshaft. It is usually made from cast iron.

It has several jobs, its main one being to smooth out the running of the engine. When the engine fires on the power stroke the piston has a tendency to boot the crankshaft to make it momentarily move quickly. Even though the engine may be running at 3500 rpm, on each power stroke it surges, then it has three dead strokes. So the flywheel absorbs the surge of the power stroke and with its stored energy keeps the crankshaft rotating through the three dead strokes.

The flywheel has fins so that as it rotates it can blow cool air over the cooling fins on the block and head to control the engine operating temperature.

It also has a magnet in it so that it becomes part of the magneto system and as such is timed to the crankshaft by a key. The starting device may also use the flywheel.

Assembling of the Engine

We now need to assemble all the parts that we have looked at to come up with an engine that would be operational.

1. Fit oil seals to crankcase and crankcase cover.
2. Fit tappets, camshaft and crankshaft (set timing marks).
3. Fit rings to piston and piston to con rod.

4. Fit piston and con rod assembly into cylinder.
5. Fit con rod cap, oil dipper and tighten con rod bolts.
6. Fit:
   • mechanical governor
   • oil slinger (if required)
   • crankcase cover and gasket.
7. Adjust crankshaft end float and tighten cover.
8. Fit valves and check tappet clearance.
9. Fit valve springs and retainers and tappet cover. Tighten.
10. Fit:
   • points
   • coil (if fitted) behind flywheel
   • flywheel and key
   • washer, nut (tighten)
   • external coil (if fitted). Adjust.
11. Check timing.
12. Fit head, gasket, and air deflector (if needed). Torque down bolts evenly.
13. Fit:
   • spark plug
   • muffler
   • carburettor and linkages
   • air filter
   • air vane governor (if fitted).
14. Fit:
   • air cowling with starter assembly
   • air baffles
   • fuel tank
   • ignition wires, switches etc.

Overhead Valve Small Petrol Engines

The overhead valve small petrol engine has been around for a long time but it is only in the last few years that manufacturers been producing them in greater numbers. In the future as pollution regulations tighten up the manufacturers of small petrol engines will have to produce cleaner burning engines, so the days of the side valve engine are numbered.

Overhead valve engines have inlet and exhaust valves in the head instead of the block. Allowing higher compression ratios causng the petrol to burn more cleanly, resulting in less pollution. As a result the engine is more efficient.

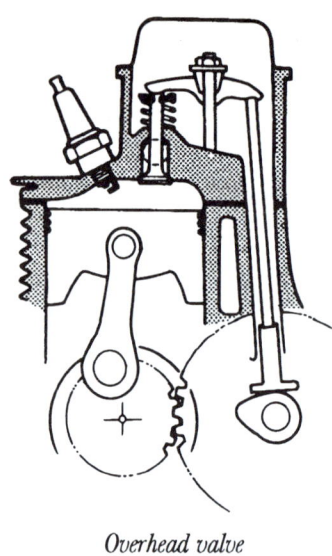

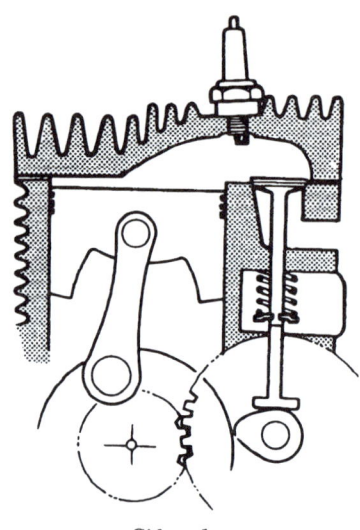

Overhead valve   *Side valve*

The valves can be opened by push rods that are operated by the camshaft situated in the cylinder block; alternatively the camshaft can be situated in the head that pushes directly onto the valve stem tip or a rocker. The overhead camshaft is usually driven by sprockets and chain.

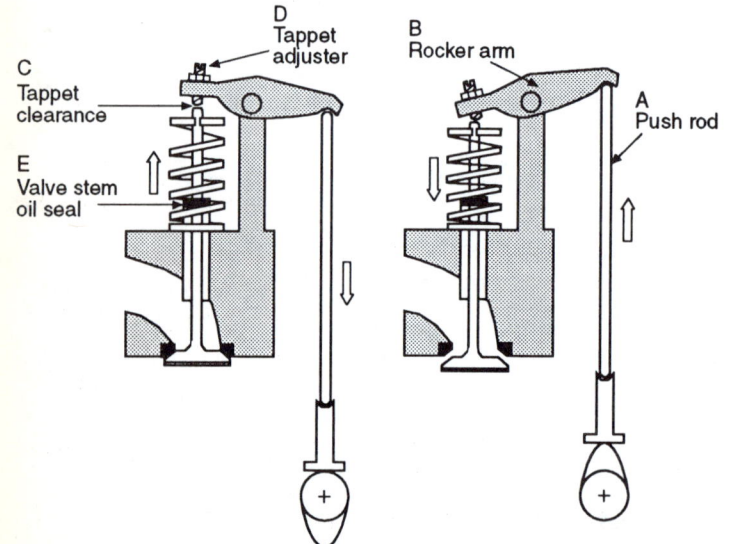

A Push rod. Tubular or solid rod that transmits reciprocating motion from tappet (cam follower) to rocker arm.

B Rocker arm. Metal arm that pivots in the centre and transfers the upward push from the push rod to downward motion to the end of the valve stem, the tip, to push open the spring loaded valve.

C Tappet clearance or valve lash. When the valve is closed there must be clearance between the valve tip and the camshaft. On an overhead valve push rod engine it is usually measured at the valve tip.

D Tappet Adjuster. A screw type adjuster for setting the tappet clearance.

E Valve stem oil seal. A seal to control the amount of oil that can lubricate the valve stem in the guide.

Operation of an overhead valve using push rods. Left, *valve closed,* right, *valve open.*

Comparison of overhead valve and side valve engines

Main features	Overhead valve	Side valve
Compression ratio	About 8.3:1	Between 6:1 and 7:1
Compression temperature and pressure	Higher	Lower
Combustion efficiency	Better	Not as good
Surface area of metal in contact with combustion area through which heat energy can be lost	Less	Higher
Heat energy lost to the cooling system	Less	Higher
Distance flame front has to travel during combusion	Short	Further
Positioning of spark plug close to hot exhaust valve to assist plug to maintain operating temperature to reduce fouling etc.	Better	Not as good
Flow of mixture into cylinder	Better	Not as good
Flow of burnt gases out of cylinder	Better	Not as good
Carbon deposits	Less	More
Fuel economy	Better	Not as good
Power output	Better	Not as good

The Two Stroke Engine

Two stroke small petrol engines have variations in their construction.

Some of the most obvious ones include:
- use of ports and reed valves instead of poppet valves
- removable barrel (cylinder) from the crankcase
- crankcase that is in two sections to enable the crankshaft to be removed
- some heads are removable while others are cast as part of the cylinder and crankcase
- crankshafts that can be dismantled
- use of anti-friction bearings instead of plain bearings
- locating pins in piston grooves to hold rings in position to stop the ring ends striking the ports and breaking the rings.

Chapter 5

Fuel System

The purpose of the fuel system is to store the petrol and supply a steady flow of petrol and air mixture as required to the engine cylinder.

Operating Principles of Carburettors

The heart of the fuel system is the carburettor. The carburettor performs three basic functions:

1. Measures the petrol and air at the right ratio for all engine operating conditions.
2. Breaks up the petrol into fine drops and mixes it with the air to form a combustible mixture.
3. Controls engine speed with a butterfly valve or throttle.

The ratio of air to petrol is approximately 15:1 by weight for normal running conditions. By volume it is about 9000:1. Although operating ratios can vary, for ease of understanding we will stick to 15:1 and talk of rich and lean mixtures.

Carburettors for small engines are simple devices compared to those that are fitted to vehicle engines. Small engines idle from 1200 to 1700 rpm and work from 3000 to 4000 rpm. Since they are either at idle or at working speed, simple carburettors can be used.

The movement of the piston down the cylinder creates a suction that draws petrol and air through the carburettor. It remains for the carburettor to mix and regulate this flow. A simple carburettor requires:

- a constant supply of petrol at a set level in a fuel bowl
- a discharge nozzle for petrol to enter the air stream
- a mixing chamber for the petrol to mix with the incoming air.

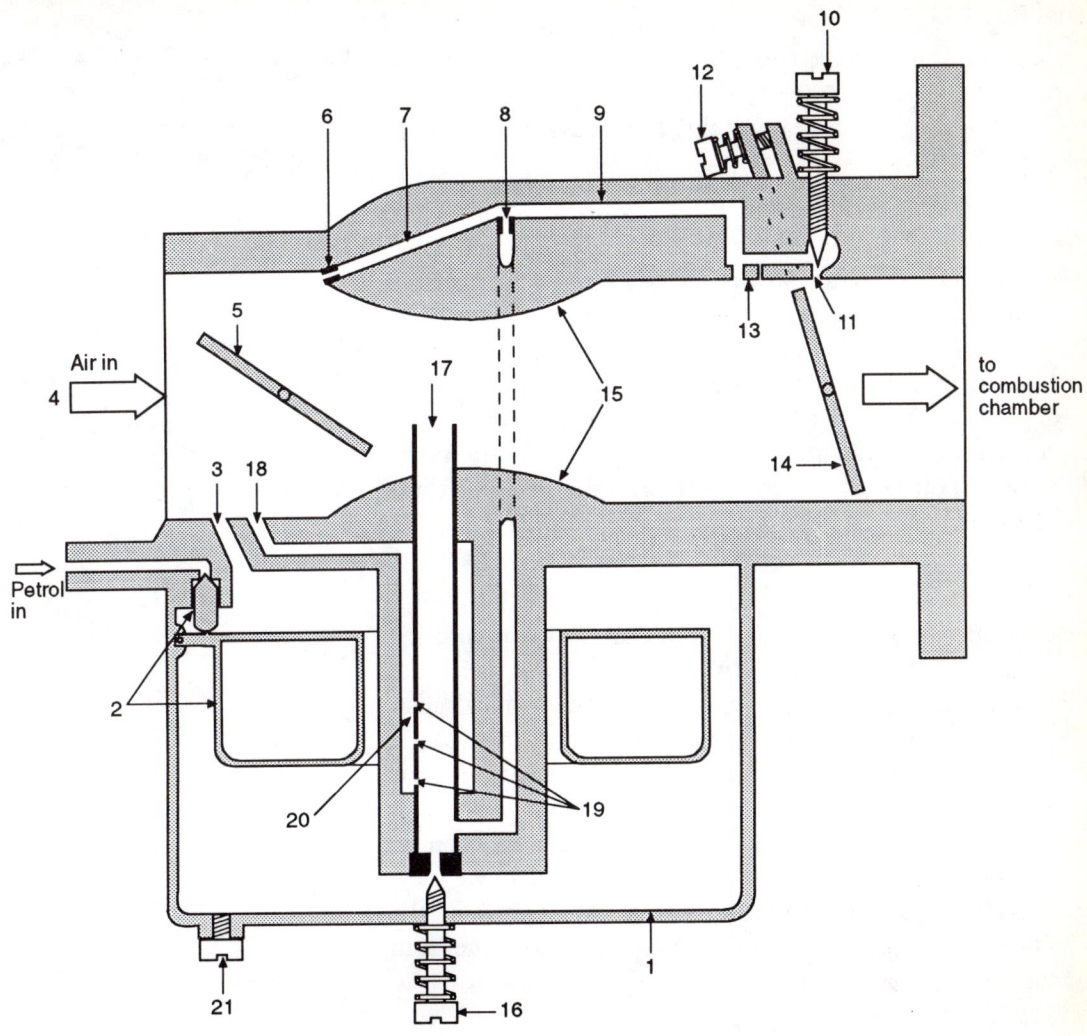

Air in

4

Petrol
in

to
combustion
chamber

1. Fuel bowl
2. Float, needle and seat
3. Internal vent for fuel bowl
4. Air intake
5. Choke
6. Idle air bleed restrictor
7. Idle air bleed passage

8. Idle fuel jet
9. Idle passage
10. Idle mixture needle valve
11. Idle discharge hole or port
12. Idle speed screw
13. Progression or transitional ports
14. Throttle butterfly valve

15. Venturi
16. Main fuel jet adjusting valve (optional)
17. Main discharge nozzle
18. High speed air bleed passage
19. High speed air bleed restrictor holes
20. Accelerator well
21. Drain plug

The parts of a basic carburettor

Fuel Bowl

To maintain a constant supply of petrol to the engine the carburettor has a fuel bowl. This is independent from the fuel tank. For ease of cleaning this bowl may have a drain plug.

Float and Needle and Seat Valve

Because the carburettor works on changes of air pressure entering the engine, it must have a constant fuel level. The level is gauged by a float in the fuel bowl, which operates a needle and seat valve to keep the fuel at the required height. If the level drops and the float falls, the needle is lowered from the seat, allowing petrol to dribble into the bowl. As the float rises it pushes the needle back onto the seat and shuts off the flow.

Venturi

The carburettor uses low air pressure to draw fuel into the air stream. To lower the pressure at one point of the intake the carburettor has a venturi. About 160 years ago an Italian scientist named Venturi discovered that if air moves past a restriction in a tube the air velocity is highest and the air pressure lowest in the area of restriction. If the air pressure in the fuel bowl is higher than that at the venturi, fuel can be drawn into the air stream. A vent into the fuel bowl leads from the beginning of the intake (before the venturi) so the fuel bowl's pressure is higher. This vent can also come from outside.

Main Discharge Nozzle

The main discharge nozzle is placed at the venturi. Fuel is drawn up the through the nozzle and fed into the air stream. The high velocity of the air breaks up the fluid into tiny droplets. This is called atomisation. The flow of fuel into the engine must be finely controlled. For this purpose a jet is placed at the base of the nozzle tube. A jet is usually a brass plug with a hole of a precise size drilled in it. This meters the fuel that flows through it.

Needle Valve

To obtain finer control of the flow of fuel through this jet, a screw with a point on it, called a needle valve, is placed at the entrance to the jet. By turning the adjusting screw in (clock-

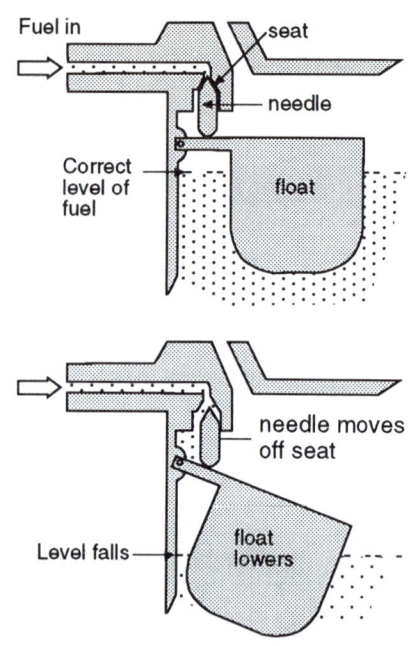

Action of float and needle and seat to control fuel level in the fuel bowl

wise) the flow is lessened (a lean mixture) and by turning it out (anti-clockwise) the flow is increased (a rich mixture).

Air Bleed

If you suck up a drink through a straw that has a hole in it, you notice air is also sucked in. This breaks up the drink as you suck. Because there is less weight of drink being drawn up, less effort is required, and the drink and air bubbles move more freely. A hole like this is an air bleed. A vent from the beginning of the intake over the carburettor delivers air around the tube to the main discharge nozzle, which has small holes in it. Air bubbles into the fuel as it is drawn in, and also helps in the atomisation process. These air bleed holes are precisely drilled to admit the required amount of air.

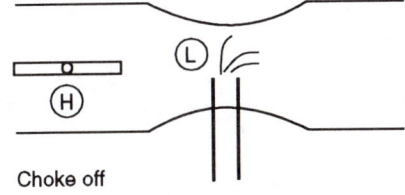

Choke off

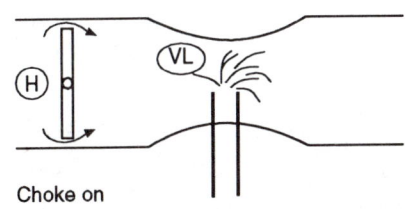

Choke on

Choke action

Choke

The choke is a butterfly valve that is placed before the venturi. When the choke is on for a cold start this valve blocks off the intake passage creating very low pressure around the main nozzle. The higher the pressure from the beginning of the intake via the vent to the fuel bowl, the more petrol is pushed into the air stream.

Throttle

The main nozzle is designed to discharge fuel into the air stream when the engine is at operational speed. In order to cater for fuel requirements at idle, further outlets and controls are needed. The control for this part of the system is the throttle. This is a second butterfly valve after the venturi. When this is open the lowest air pressure is in the venturi. When the throttle is closed it blocks off the air flow before it, and the low air pressure then occurs after the valve, and around its edges. The extent to which the throttle can close is regulated by the idle speed screw.

Idle System

When the engine is at idle and the throttle closed, no fuel is drawn out the main discharge nozzle as the air pressure before the throttle valve is not low enough. A passage from below the discharge nozzle takes fuel up to discharge after the throttle valve. The passage is fitted with another precisely drilled jet to regulate the flow, and an adjustable needle valve is fitted to the discharge port. Another passage acts as an air bleed to the idle passage. This too is fitted with a restricting jet.

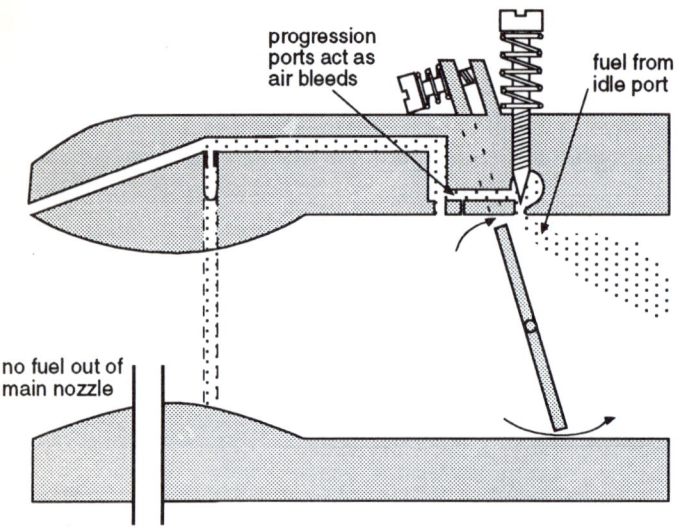

progression
ports act as
air bleeds

fuel from
idle port

no fuel out of
main nozzle

Engine at idle speed (throttle closed)

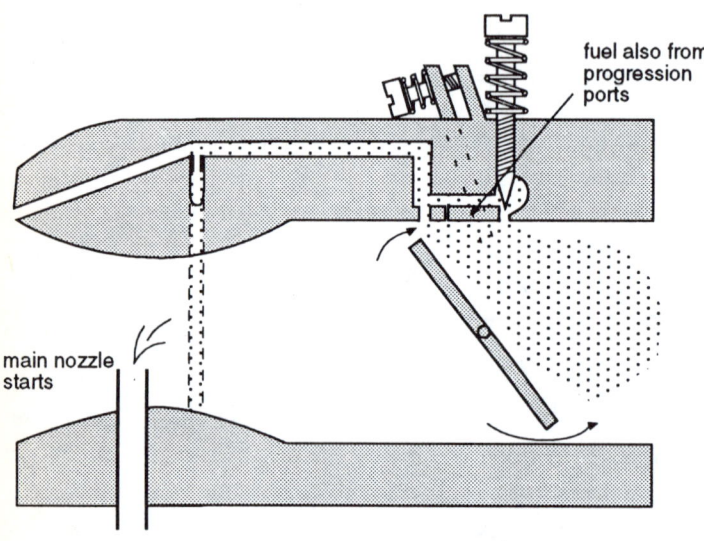

fuel also from
progression
ports

main nozzle
starts

Engine moving above idle speed (throttle opening past progression ports)

Progression or Transitional Ports

When a small engine changes from idle to working speed, about 3500 rpm, the carburettor changes from the idle system to the main system, so that the engine draws petrol from the main discharge nozzle. A problem then arises that air responds quicker to pressure change than petrol. As the throttle valve is quickly opened, the engine momentarily draws in too much air before sufficient petrol is drawn out of the main discharge nozzle at the correct ratio. Manufacturers use different methods of reducing the lean spot when the engine is changing from idle to working speed.

Some methods are the use of an automatic choke that responds to the engine's needs, to momentarily close, in order to enrich the mixture. A more common method is the use of progression ports in the idle system.

A series of ports is drilled in the idle passage back towards the venturi. As the throttle valve moves off the idle position, progressively more petrol can be drawn into the carburettor throat as the low pressure beneath the throttle valve moves along the throat.

Other devices are used such as a cavity well at the progression ports to hold extra petrol. Progression ports also act as air bleeds when the engine is at idle.

Accelerator Well

In a similar way to the changing needs of the engine from idle to full throttle, fluctuations in the fuel required occur during operation, due to changes in load or throttle adjustments. To cope with moving from idle and these other changes, a carburettor may have an accelerator well around the main discharge nozzle.

The discharge nozzle has a series of small holes in the lower half. The holes serve two purposes. One is to let the petrol move freely between the inside of the nozzle and the well. The other is to act as air bleeds.

On accelerating, the throttle valve is opened wide and the low pressure from below the throttle valve changes to the venturi. The nozzle now has extra petrol in the well to quickly discharge into the venturi. The low pressure in the venturi draws out the petrol and the level in the well quickly drops to supply a rich mixture to the engine as it changes over from the idle to operational speed, thus eliminating the "miss".

When the engine reaches operating speed and is under full load with wide open throttle, the accelerator well will be empty and the petrol is then entering from the main jet. The accelerator well will refill when the throttle valve drops back to idle or about one quarter throttle, or when the engine stops. Once refilled the accelerator well can then respond to the engine's need for extra petrol as required.

The action of air bleeds and nozzles discharging into flowing air atomises the fuel into tiny droplets. Low pressure also helps this. Because petrol is a volatile fuel it also vaporises easily, that is, becomes gaseous. Low pressure and the heat

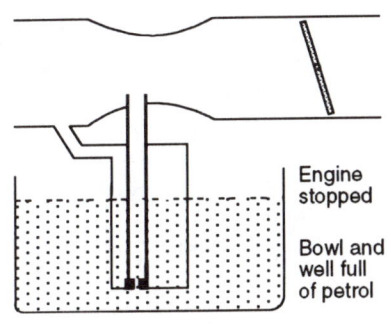

Engine
stopped

Bowl and
well full
of petrol

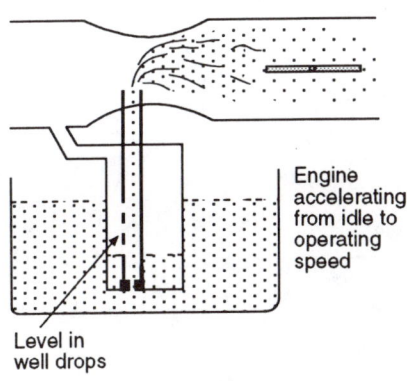

Engine
accelerating
from idle to
operating
speed

Level in
well drops

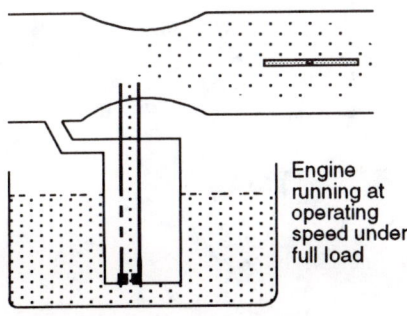

Engine
running at
operating
speed under
full load

The action of an accelerator well

from the engine help this. As the petrol/air mixture leaves the carburettor the passage narrows yet again. This reduces pressure further and assists in final vaporisation. The carburettor's task is now complete.

We will now look briefly at some other types of carburettors that are used on small engines.

Airfoil Carburettors

An airfoil is a device that creates a difference in pressure when air passes over it. An aeroplane wing is an example of an airfoil. The airfoil used in some carburettors is like a piece of pipe. It is usually found in the venturi area where the air pressure is low.

Using an Airfoil

The airfoil also acts as a passage to move the petrol to an idle discharge port and the hole in it then becomes an air bleed. When the choke is being used petrol will be drawn out of the airfoil discharge hole and idle ports. An air bleed can be incorporated into the lower end of the airfoil tube.

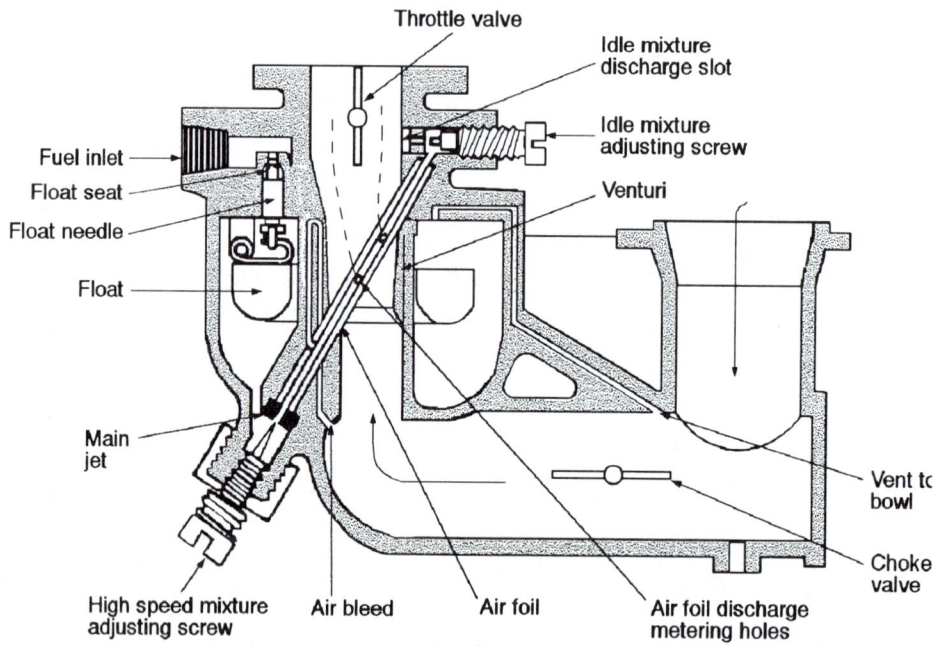

Gravity feed updraught float type carburettor using an airfoil. Engine running at operating speed under full load.

Using a Long High Speed Mixture Adjusting Screw

This carburettor uses a discharge nozzle with a high speed mixture adjusting screw needle coming in from the top of the carburettor down through the nozzle to the jet at the bottom. The needle is hollow to allow petrol to be drawn up the tube. Near the point of the needle is a small hole to let the petrol in and a hole near the top to let the petrol and air mixture into the idle passage. There is also a hole in the tube in the venturi area which acts as an idle air bleed.

Suction or Vacuum Feed Carburettors

Suction feed carburettors are of simple design and construction. They are attached to the top of the fuel tank so no fuel bowl is required. A suction tube goes down from the carburettor to near the bottom of the tank. At the base of the tube is a fine fuel filter screen and above the filter is a check ball to stop the petrol draining back down the tube.

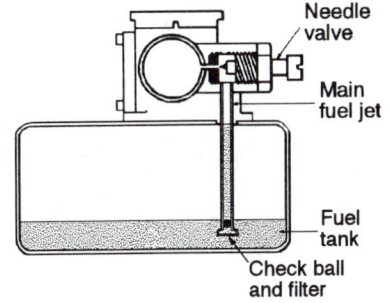

The fuel tank is vented by means of a hole in the cap. If this hole becomes blocked the atmospheric pressure will not be able to enter the tank to push down on the top of the petrol as the fuel is used, so the engine will stop.

When the engine is at idle the carburettor acts like the carburettors we have been looking at.

The low pressure below the throttle valve at idle draws petrol out of the idle discharge hole. There is only one mixture adjusting screw so the idle discharge hole is of a precise diameter to meter the petrol. The high speed discharge hole will also act like an air bleed at idle. With the engine at idle the atmospheric pressure enters the fuel tank through the vent hole in the cap to push the petrol up the suction tube through the jet, then out through the idle discharge hole into the low pressure area created by the intake stroke.

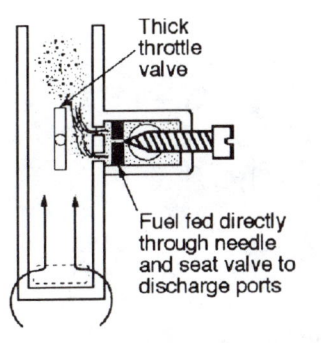

Suction feed carburettor side and top view

When the intake stroke is finished the area of low pressure below the throttle valve ceases and the petrol in the suction tube wants to drop down the tube back into the tank but is prevented from doing so by the ball check valve at the base of the tube. The ball check valve ensures the petrol is always available to the top of the tube at the adjusting screw and jet.

When the throttle valve is opened both the idle and high speed discharge holes are brought into use. There is no round type of venturi in this style of carburettor. So to maintain low pressure in the carburettor a restriction is placed at the choke.

Also the throttle valve is thicker than normal, so the air flow has to speed up as it goes around the open throttle valve and as a result, we have a venturi effect and petrol is drawn out of both the discharge holes. The amount of petrol entering the carburettor throat at operating speed is controlled by the mixture adjusting screw.

When the engine returns to idle speed the low pressure below the throttle valve will only draw petrol out of the idle discharge hole, so petrol ceases to be drawn out of the high speed discharge hole.

To allow for acceleration the mixture is set slightly rich. As the petrol level in the tank drops the engine will run lean towards the bottom of the tank as compared to running rich when the tank is full. The carburettor is usually adjusted when the tank is a quarter to half full to allow for the changes in the mixture ratio as the fuel level alters.

To get an engine started when cold with this type of carburettor the choke is usually closed tight to aid the suction effect.

In some types of suction carburettors a spiral may be fitted between the throttle valve and inlet valve to assist the operation of the carburettor. The two discharge holes are of a precise diameter.

Suction Feed Carburettor Using a Fuel Cup

To avoid a lean mixture when the fuel tank is nearly empty and a richer mixture when the tank is full, a fuel cup or bowl is fitted to this type of carburettor to allow a constant air/petrol ratio regardless of the petrol level in the tank.

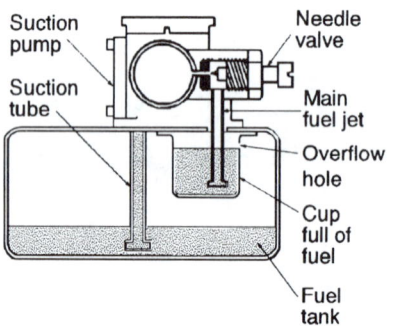

Suction feed carburettor using a fuel cup and pump

A fuel pump is used to fill the fuel cup. The fuel tank, fuel pump and fuel cup serve the same functions as the gravity feed tank, the fuel bowl, float and needle and seat of the float type carburettor. A diaphragm type of pump is used that works off the pressure fluctuation in the inlet pipe between the throttle valve and the inlet valve.

When the engine inlet valve opens the low pressure pulls the pump diaphragm against a light coil spring. The flexing diaphragm then draws petrol out of the tank through the open inlet valve of the pump and into the fuel pump chamber. When the engine inlet valve closes the low pressure ceases, the light coil spring that was compressed then releases its energy and pushes against the flexed diaphragm and the petrol is then forced out of the discharge valve to the fuel cup.

The pump action keeps going all the time. When the fuel cup is full the excess petrol is dumped out an overflow hole back into the tank. Check valves are not generally used in the long suction tube, because the valves in the fuel pump perform the same function.

Maintenance of Carburettors

The carburettor usually gives reliable service until something like dirty fuel gets into the system or the engine has been left standing for a long time and has gummed up passages and jets. Vibration is also a problem and parts come loose.

There are additives available to keep petrol fresh for extended periods which some engine manufacturers recommend. If the engine is stored with petrol in it these will stop gumming. Engines using carburettors with diaphragms should not be stored with fuel still in them.

There are some general rules applicable to all carburettors:
*   Neither petrol nor air must ever leak from any part of a carburettor.
*   Every opening so intended must always pass exactly the amount of petrol or air intended by the manufacturer.

The drawings in the previous section naturally show passages and jets a lot larger than in the real carburettor. The manufacturer has worked out the size of passages and jets and these sizes must not be altered. They should be kept clean and all surfaces and gaskets should be in good condition.

The carburettor is made of many pieces. These pieces must fit together correctly and be kept tight and adjusted correctly.

Basic Carburettor Adjustment

Most carburettors have some type of adjustment. Some later carburettors have less adjustments than older ones.

Float Type Carburettor
Basic adjustments cover:
*   idle speed
*   idle mixture
*   high speed mixture.

Some carburettors have a non-adjustable main jet. It is a precisely drilled hole to control the flow of petrol to suit the engine's petrol requirements under all operating conditions.

The following is a general adjustment procedure for a four stroke small engine carburettor. (Always refer to the manufacturer's instructions for exact details.)

1. Initial settings, if needed, of idle and high speed mixture valves:
- Idle valve: Gently turn in until it just closes, then turn out 1 1/2 turns.
- High speed valve: Gently turn in till it just closes then turn out 1 1/2 turns.

Valves may be damaged if they are turned in too far.

2. Engine running at operating temperature, air cleaner fitted.
- Place the throttle control lever to full operating speed.
- Turn the high speed needle valve in until the engine slows (clockwise, lean mixture).
- Turn it out past smooth operating point (rich mixture).
- Set the high speed needle valve to the midpoint between rich and lean.
- Place the throttle control lever at idle speed.
- Place your hand on the throttle lever at the carburettor to hold it against the idle screw and adjust the idle speed.
- Whilst still holding the throttle lever, adjust the idle mixture needle valve in (lean) and out (rich) and set at the midpoint between rich and lean.
- Recheck idle speed then release the throttle lever.
- Check if the engine accelerates smoothly, if not readjust the carburettor to a slightly richer mixture.

Suction Feed Carburettors

If the suction feed carburettor is not fitted with a fuel cup the tank should be about one quarter full. If the carburettor has a fuel cup the tank should be half full. With these levels you get the best adjustment.

1. Initial setting, if needed, of needle valve.
- Gently turn in until it just closes then turn out 1 1/2 turns.

2. Engine at operating temperature with air cleaner fitted.
- Place throttle control lever to full operating speed.
- Turn the needle valve in until the engine slows (clockwise, lean mixture).
- Turn the valve slowly out until the engine runs smoothly then take the valve out a little bit more until it begins to run unevenly (slightly rich). Since this setting is made without load, the mixture setting should operate the engine satisfactorily under load.

- Place the throttle control in the idle position.
- Rotate the throttle lever on the carburettor to hold against the idle throttle stop and then adjust the idle speed screw to obtain the desired idle speed.
- Move throttle control from idle to full operating speed and the engine should accelerate smoothly; if not, adjust the needle valve to give a slightly richer mixture.

Adjusting a Manual Choke
These are chokes using a wire cable:
- Place the choke control to the fully closed position.
- Loosen the screw securing the cable wire (located either at the control lever or at the carburettor).
- With the choke completely closed tighten the wire clamp screw.
- Check choke operation to see that it fully opens and closes.

Diaphragm Carburettor as used on Chainsaws
- Screw both low and high speed mixture jets in until they are lightly seated.
- Turn both jets out one full turn (this is the basic setting; otherwise refer to manufacturer's handbook).
- Start the saw and allow it to warm up.
- Adjust the low speed jet first to achieve the smoothest idle possible.
  - Enrich the mixture (turn out) if there is hesitation when snap accelerating (from idle to full throttle).
  - Reduce the mixture (turn in) if the saw runs roughly or smokes at idle.
  - Only turn 1/10 revolution at a time.
- Readjust the throttle stop screw so the engine is idling smoothly without the chain running.
- High speed jet
  - Enrich – turn out 1/4 turn.
  - Full throttle (no load) – the engine should "4 stroke".
  - Test cut at maximum revs – the engine should "2 stroke".
  - Adjust the high speed jet – lean (turn in) 1/10 turn at a time until maximum power is achieved.
  - The high speed jet must not be reduced more than 7/8 (out) overall.
- To confirm maximum tuning:
  - The saw should "4 stroke" at maximum revs with no load and
  - "2 stroke" at maximum revs with full load, to give maximum performance.

- After low and high speed jets are correctly adjusted, it may be necessary to readjust the throttle stop screw to adjust the idle speed.

> Use a well fitting small screw driver
> to rotate mixture screws.

Disassembly, Cleaning and Adjustment

If the carburettor is gummed up from not being started for many months it may need cleaning out. If you are to do it yourself, it pays to have a workshop manual. If the carburettor is dismantled you should take note of settings and adjustments prior to the operation. Lay the parts out in a logical order on a clean surface. Use suitable cleaning solvents and compressed air. Ensure all passages and jets are clear and clean. Check surfaces for flatness. Parts must be spotless before reassembly. A carburettor overhaul kit should be used where possible. Never use gasket sealants on carburettor gaskets because the jets and passages are easily blocked.

We will now look at how you dismantle and reassemble two types of carburettors.

Only dismantle a carburettor if you have the correct tools and good screwdrivers, the right instructions and the right mechanical knowledge.

The first carburettor is from Honda GX120KI and GX160KI engines and is fairly typical of carburettors used on many Honda engines.

The second carburettor is a Briggs and Stratton Two Piece Flo-Jet type, found on many of their engines.

Honda Carburettor

Note: CAUTION when working with petrol.

Disassembly

1. Clean the carburettor and its surroundings with suitable solvent and compressed air before you start to remove it.

2. Remove the air cleaner assembly.

3. Remove the high tension lead from the spark plug.

4. Lift up the choke lever and remove it from the carburettor.

5. Disconnect the fuel hose from the carburettor. Use the short peg on the choke lever to plug the end of the fuel hose.

5. Remove the throttle return spring.

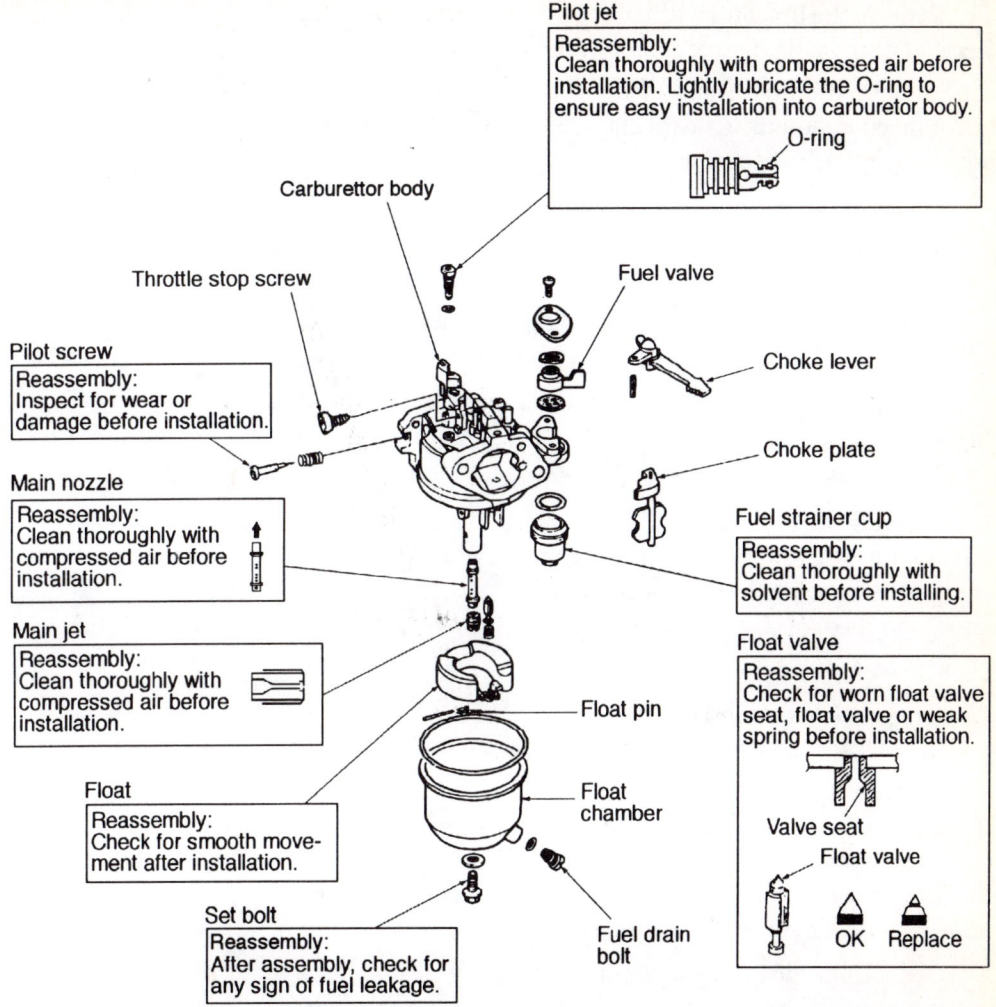

Pilot jet
Reassembly:
Clean thoroughly with compressed air before installation. Lightly lubricate the O-ring to ensure easy installation into carburetor body.
O-ring

Carburettor body

Throttle stop screw

Fuel valve

Choke lever

Pilot screw
Reassembly:
Inspect for wear or damage before installation.

Choke plate

Main nozzle
Reassembly:
Clean thoroughly with compressed air before installation.

Fuel strainer cup
Reassembly:
Clean thoroughly with solvent before installing.

Main jet
Reassembly:
Clean thoroughly with compressed air before installation.

Float valve
Reassembly:
Check for worn float valve seat, float valve or weak spring before installation.

Float pin

Valve seat

Float chamber

Float valve

Float
Reassembly:
Check for smooth movement after installation.

OK   Replace

Set bolt
Reassembly:
After assembly, check for any sign of fuel leakage.

Fuel drain bolt

7. Disconnect the governor rod by pulling the carburettor forward to a point where the groove in the throttle arm lines up with the rod, and lift the rod out of its hole. (Reassembly is in the reverse order of disassembly.)

8. Remove the carburettor, gasket and insulator.

9. Before proceeding make sure you have a clean area on which to work and lay out the parts in a logical order as shown in the exploded view of the carburettor. Make sure you are using metric tools and well fitting screwdrivers. Also ensure the outside of the carburettor is clean before you operate.

10. Remove the drain bolt and drain the carburettor before dismantling it. Fuel vapour or spilled fuel may ignite.

11. Remove the bolt and washer at the base of the float chamber and remove the float chamber and its sealing ring.

12. Remove the float pin and float, together with the float valve (needle) and its light spring. Keep an eye on that small spring or it could disappear forever.

13. Remove the main jet with a well fitting screwdriver and then the main nozzle will drop out.

14. Remove the throttle stop screw (idle speed adjusting screw).

15. Carefully remove the pilot jet (idle jet) by levering it up out of its hole. It is plastic material fitted with a small O ring.

16. Remove the pilot screw (idle mixture adjusting valve) with a well fitting screwdriver.

17. Unscrew the fuel strainer cup and remove the O ring.

18. Unscrew the two retaining screws, holding the fuel valve in with a well fitting screwdriver, and remove the fuel valve components, noting how they are fitted.

The carburettor is now dismantled as far as it needs to go for a good clean out.

Cleaning and Inspection

1. Before cleaning the parts look for obvious blockages, wear or damage.

If water is allowed to accumulate in the bottom of the float chamber over a period of time, it can cause the bolt that holds the float chamber on to react with the aluminium carburettor housing and it may refuse to undo.

2. WARNING: To prevent serious eye injury, always wear goggles or other eye protection when using compressed air.

3. CAUTION: Some commercially available chemical cleaners are very caustic. These cleaners may damage plastic parts such as O rings, floats and float valve seats. Check the container for instructions. If you are in doubt do not use these products to clean Honda carburettors. High air pressure may damage the carburettor. Use low pressure settings when cleaning passages and parts.

4. Clean the carburettor body and parts with suitable solvents. Do not use water. Make sure all gasket surfaces are absolutely clean of all gasket material, including where the carburettor attaches to the engine.

5. Use low air pressure and clean the following parts and passages:
• internal or external vent port

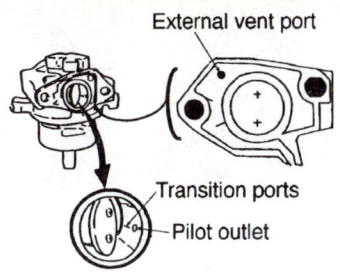

Engine block side

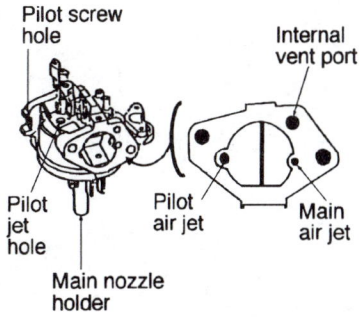

Air cleaner side

- pilot screw hole
- pilot jet hole
- pilot air jet
- pilot jet
- pilot outlet
- transition parts
- main nozzle holder
- main nozzle
- main jet
- float valve seat
- fuel passage into carburettor
- carburettor body and associated parts.

NOTE: Both internal and external vent ports are found on the carburettor. In carburettors that are externally vented, the external vent passage is open to the carburettor bowl and the internal passage is closed. In carburettors that are internally vented, the internal passage is open to the carburettor bowl and the external vent passage is closed.

6. Check all parts for damage or wear. Note the condition of the float valve where it seats and the pointed end of the pilot screw. Replace parts that are suspect.

7. Check that the throttle and choke valves and shafts are satisfactory and tight.

8. Inspect the float level height.

Reassembly

1. Always fit new gaskets, O rings and sealing washers when reassembling the carburettor. Do not use gasket sealants on carburettor gaskets because the jets, passages and drillings etc. are easily blocked.

2. Reassemble the carburettor in the reverse order of disassembly.

3. Screw in the pilot screw until it lightly bottoms out and screw it out the necessary number of turns as listed under carburettor adjustment. The pilot screw will receive its final adjustment when the engine is warmed up.

4. Screw in the throttle stop screw until the throttle valve just begins to open, then give the screw half a turn more, so that the throttle valve is slightly open. This base setting should be enough to get the engine running, then the idle speed can be correctly adjusted.

5. Fit the carburettor to the engine with the gaskets and

insulator positioned correctly. Check that the governor rod and spring are correctly fitted and free to move.

6. With the fuel line connected and adequate petrol in the tank, check for any petrol leaks at the carburettor. (When a carburettor has had a good clean out, the fuel filters and tank are usually checked and cleaned if required.)

7. Refit the air cleaner. Check over your work to make sure everything is in its place and secure.

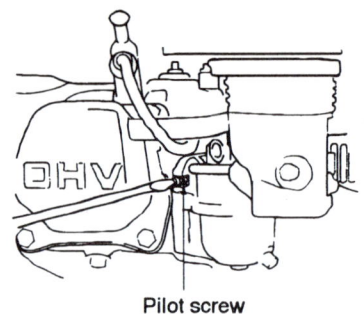

Pilot screw

Carburettor adjustment

1. Start the engine and allow it to warm up to normal operating temperature.

2. With the engine idling, turn the pilot screw in or out to the setting that produces the highest idle speed. This will have to ascertained by trial and error for any given engine.

3. After the pilot screw is correctly adjusted, turn the throttle stop screw to obtain the standard idle speed, which is usually about 1400 rpm.

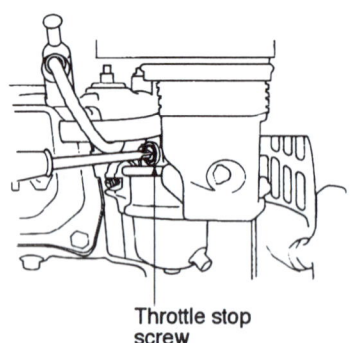

Throttle stop screw

Briggs & Stratton carburettor

Note: CAUTION when working with petrol. Make sure that the ignition is switched off or the high tension lead is earthed out before any maintenance work on the engine is commenced.

Disassembly

1. Clean the carburettor and its surroundings with a suitable solvent and compressed air before you start to remove it.

2. Remove the air cleaner assembly.

3. Remove the two screws that hold the carburettor onto the inlet pipe and carefully rotate the carburettor to unhook the governor rod.

4. Check that the outside of the carburettor is clean before you operate.

5. Before proceeding make sure you have a clean area on which to work and lay out the parts in a logical order. Make sure you are using A.F. spanners and well fitting screwdrivers.

6. Remove the idle mixture valve with a well fitting screwdriver.

7. Remove the packing nut and the high speed needle valve together.

8. With a well fitting screwdriver remove the nozzle. (Be careful

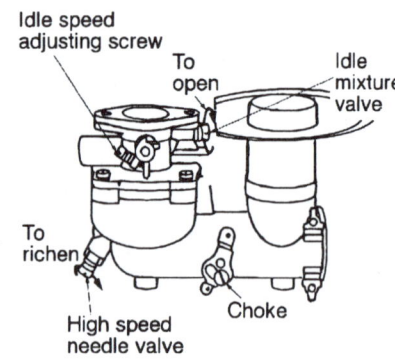

Idle speed adjusting screw / To open / Idle mixture valve / To richen / High speed needle valve / Choke

not to damage the internal threads. If they are damaged a special tap is available to repair them).

Because the nozzle projects into the top half of the body, it must be removed or it could be bent when the two parts are separated.

9. Remove the three screws, with a well fitting screwdriver, and the upper and lower bodies will separate.

10. Remove the float pivot pin and remove the float and float needle as an assembly. Remove the gasket. The float needle can be removed from the float.

The carburettor is now dismantled as far as it needs to go for a good clean out.

Cleaning and inspection

1. Before cleaning the parts look for obvious blockages, wear or damage.

2. WARNING: To prevent serious eye injury, always wear goggles or other eye protection when using compressed air.

3. Clean the carburettor and parts with suitable solvents. Do not use water. Make sure all gasket surfaces are absolutely clean of all gasket material, including where the carburettor attaches to the inlet pipe.

4. Use low air pressure and clean the following parts and passages:
- nozzle
- nozzle passage
- vent passage to float chamber
- vent passage to nozzle
- idle mixture valve hole and discharge hole/slot
- fuel inlet passage
- inside and outside of carburettor
- all associated parts.

5. Check all parts for damage or wear. Note the condition of the float needle where it seats, also that the float does not have a hole in it. Check the pointed ends of the two adjusting valves. Check that the throttle and choke butterfly valves and shafts are clear and tight. Replace all damaged, worn or suspect parts.

If the throttle shaft is suspect it can be checked to see if the wear is greater than the 0.25 mm (.010") allowed. Check wear by placing a short iron bar on the upper body as shown. Measure the distance between bar and shaft with a feeler gauge

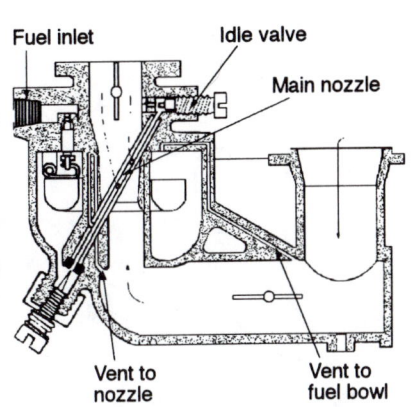

Fuel inlet   Idle valve

Main nozzle

Vent to nozzle   Vent to fuel bowl

Parts to be cleaned with compressed air

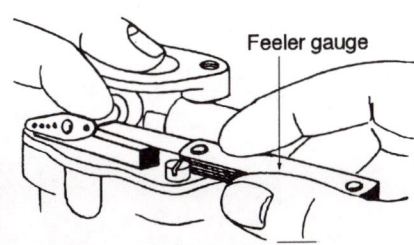

Feeler gauge

Checking throttle wear

while holding the shaft down then holding shaft up. If the difference is over 0.25mm, either the upper body should be rebrushed, the throttle shaft replaced, or both. Wear on the throttle shaft can be checked by comparing worn and unworn portions of the shaft.

6. If the upper carburettor body is suspected of being warped, it can be checked by the use of a 0.05mm (.002") feeler gauge.

With the carburettor top and bottom screwed together with a new gasket, if the 0.05 mm feeler gauge can be inserted between the upper and lower bodies, just below the idle valve, the upper body is warped or gasket surfaces are damaged and should be replaced.

7. Inspect the float level height. Firstly refit the needle to the float, then refit the float after the body gasket is fitted.

Turn the body and float upside down and the float should be parallel to the body mounting surface. If not, bend the tong on the float until they are parallel. Do not press on the float.

Reassembly
1. Always fit new gaskets when reassembling the carburettor. Do not use gasket sealants on carburettor gaskets because the jets, passages and drilling are easily blocked.

2. Reassemble in the reverse order of disassembly.

3. Screw in the idle mixture valve until it lightly bottoms then turn it 1 1/4 turns out.

4. Screw in the high speed needle valve until it lightly bottoms, then turn it 1 1/2 turns out.

5. If the idle speed screw has not been touched, it should be correct. If it has been, turn the screw until the valve starts to open from fully closed, then give it half a turn to open the valve slightly, which should be enough to use as a base adjustment.

6. Fit the carburettor to the engine with a new gasket. Check that the governor rod is correctly fitted and free to move.

7. Refit the air cleaner. Check over your work and make sure that everything is in its place and secure.

When a carburettor has had a good clean out, the fuel filters and tank are usually checked and cleaned if required.

Carburettor adjustment
1. Start the engine and allow it to warm up to its normal operating temperature.

2. Place the governor speed control lever in the fast position. Turn the high speed needle valve in until engine slows (clockwise – lean mixture). Then turn it out past smooth operating point (rich mixture). Now turn the high speed needle valve to the midpoint between rich and lean.

3. Next, adjust the idle mixture. Holding the throttle against the idle stop, turn the idle midway between rich and lean. Recheck idle speed. Release the throttle. If the engine will not accelerate properly, the carburettor should be readjusted, usually to a slightly richer mixture.

Fuel Tanks, Lines and Filters

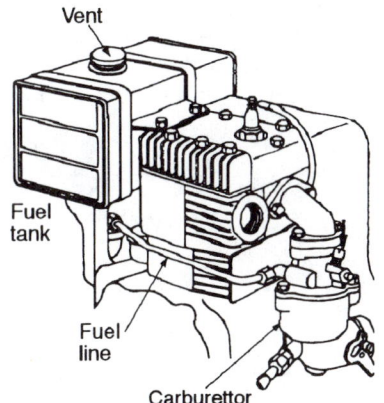

The carburettor is the most complex component of the fuel system, but the other parts all serve a purpose and must be looked after.

Gravity feed carburettors have the fuel tank mounted above the level of the petrol in the fuel bowl so that the petrol is free to flow from the tank to the carburettor. A tap is usually on the outlet and a vent hole is in the fuel cap. If the vent hole in the cap is blocked fuel will cease to flow out the open tap. It is important to keep the vent hole clear.

Suction feed carburettors have the tank mounted under the carburettor, also with a vented cap.

Fuel tanks and lines are usually made from steel or special plastics.

The fuel is usually filtered as it leaves the tank. The fine mesh filters are in the tank on smaller engines and on larger ones a glass bowl and strainer are fitted. The glass bowl allows water and dirt to drop to the bottom of the bowl as the petrol leaves the tank. Suction carburettors usually have a fine mesh screen on the bottom of the suction tubes. Some small inline filters are used, similar to those used on vehicle, bike and boat engines.

Maintenance of Tanks, Lines and Filters

It is very important that the tank, lines and filters are kept clean. Dirt and water in the fuel system is a problem to the carburettor as the small openings are easily blocked. Petrol flows more freely through a small hole then does water.

If the tank has rust or dirt in it, it should be taken off and thoroughly cleaned.

The lines should always have a good flow of petrol through them. If the line is clear and the flow is restricted it could be due to a blockage of the vent in the cap, the tap or a filter.

Filters that are in the tank of a gravity feed system can be cleaned by removing the tap from the tank. Systems using glass bowls at the tanks can be serviced by removing the bowl and gauze screen.

On suction feed carburettors the carburettor has to be removed from the tank to clean the fine mesh screens.

Air Cleaners

The air cleaner, or filter, is a simple but very important part of the engine. Its function is to filter the dirt, dust and other foreign particles out of the air as it enters the carburettor. A filter removes all but the minutest particles out of the air. If the air cleaner were left off a small engine in a dusty paddock the dirt and dust entering the cylinder would rapidly wear the internal moving parts of the engine. When the cleaner becomes partially blocked, the engine performance drops so the cleaner must be serviced as required, according to the manufacturer's specifications.

Three types of filters are generally found on small engines:
* oil bath air cleaners
* foam rubber cleaners
* dry cartridge cleaners.

Oil Bath Air Cleaners

These are usually found on older engines. The air entering the cleaner has to go down then turn around to go up through a fine mesh gauze. As it turns above the oil, the heavy particles hit and are trapped by the oil. The air then proceeds up through the oil wetted fine mesh gauze for a final clean before entering the carburettor.

Foam Rubber Cleaners

The foam rubber cleaner is usually wet from oil to help trap the dirt and dust as it passes through the foam.

Dry Cartridge Cleaners

Dry cartridge air cleaners are similar to those used in vehicles and tractors. They are a very efficient means of cleaning the air. Some dry cartridges use a foam rubber precleaner.

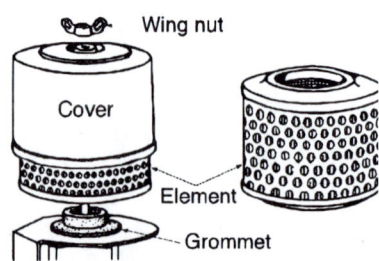

Dry cartridge air cleaner

Governors

The purpose of the governor on a small petrol engine is to maintain, within certain limits, a desired engine speed, even though the load may vary.

The governor is an automatic device that opens the throttle valve to allow more petrol/air mixture to enter the cylinder when the work load gets heavy and closes the throttle valve when the work load is light. Thus the engine speed changes very little no matter what the load.

The governor also stops the engine from over revving. When an engine revs over its recommended maximum operating speed mechanical damage usually results.

Most of the small engines in use have a throttle control whereby the operator can set the engine to a desired speed, such as the speed required for working a water pump or auger. Some engines have a set speed governor, so that when the engine starts up, it revs out to that speed, as on some generators.

The four stroke small petrol engine usually has an air vane or mechanical type governor fitted to it.

The Mechanical Governor

This is also known as a centrifugal or counterbalance governor. It uses a set of hinged weights to work against the pull of a spring attached to the throttle. The weights are attached to a gear on a shaft that is rotated by the crankshaft or camshaft. As the engine revs, the flyweights will move outwards by centrifugal force. If the speed is high enough, the weights will move out to their maximum. As the engine revs decrease, the flyweights will drop back because they are not spinning as quickly.

The weights work against a plunger and lever system to control the throttle. The hand throttle control is not hooked directly to the throttle valve on the carburettor; it is hooked up using a extension spring.

When the engine is stopped with the hand throttle control set at the idle position, the slight tension in the governor spring usually holds the throttle valve at the carburettor in the full open position.

With the engine at idle the slight tension of the governor spring is opposed by the centrifugal force of the flyweights and the plunger is moved away to push against the crank,

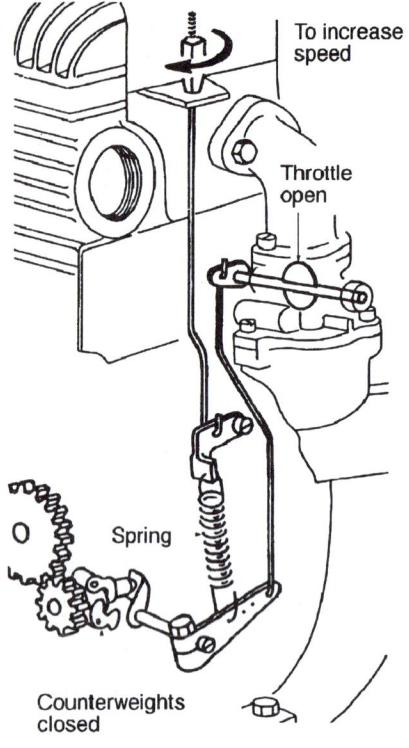

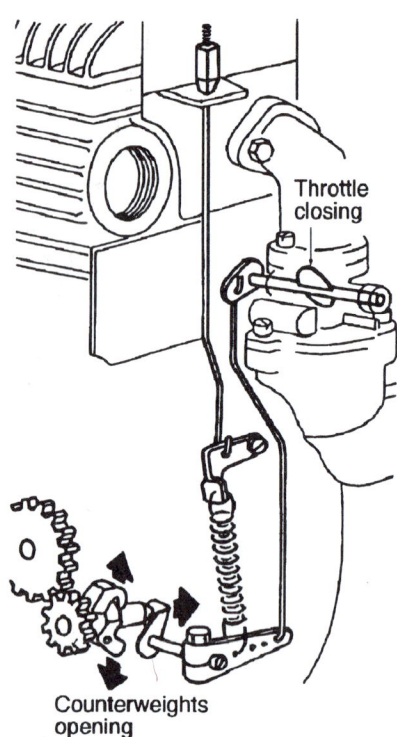

which moves the throttle lever against the idle speed screw stop. This keeps the engine at idle.

When the hand throttle is moved to the operating position with no load on the engine, the governor spring stretches to open the throttle valve wide. But the centrifugal force of the quickly spinning flyweights closes off the throttle valve so that only enough mixture can enter the cylinder to keep the engine running at no load revs, in this case 3700 rpm.

When the load is applied, the engine speed will drop slightly and then the centrifugal force of the flyweights will lessen. The stretched governor spring would then be unopposed and pull the throttle open, maintaining the speed and giving the engine the required amount of fuel mixture.

If the load were removed, the engine would rev more quickly, but the governor would react immediately to hold the engine to the no load revs.

When a small engine is operating at speed the engine does not always operate at maximum load, so the throttle valve in the carburettor moves between part throttle and almost fully open throttle.

If the engine revs increase the weights are forced out, pushing against the crank to close the throttle

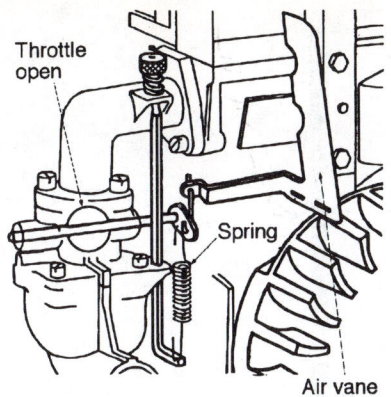 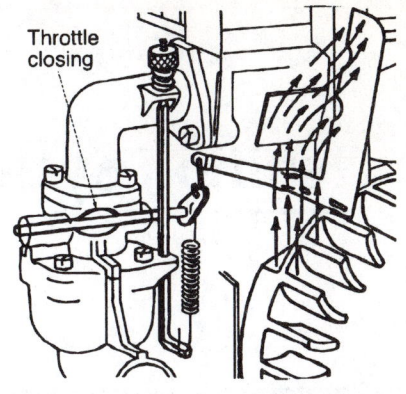

Air vane

The Air Vane Governor

The air vane or pneumatic governor uses the air pressure from the flywheel fins to push against a vane to provide the force to oppose the force of the governor spring.

The air pressure from the flywheel fins is proportional to engine speed. The faster the engine spins the greater the force of air on the vane lever.

General Points

The engine manufacturers make governors matched to suit their engines. Most governors have some type of adjustment, such as moving the governor spring or throttle lever link to another hole. These types of adjustment should not be carried out unless necessary, for tampering with the governor may cause the engine to overspeed.

Maintenance

* All pivot points and connections should be free to move, or excess friction will cause the governor to malfunction.
* Badly worn and loose parts will cause malfunctions.
* A tachometer should be used to check engine idle and no load top speed.
* To correctly adjust the governor refer to the engine manufacturer's instructions.

Chapter 6

Ignition System

The purpose of the ignition system is to provide a spark in the combustion chamber at the correct time to ignite the petrol/air mixture.

The flywheel magneto is the most common type of small engine ignition system and that is the type we will be looking at. A magneto is an electrical device that provides ignition to an internal combustion engine without the aid of a battery.

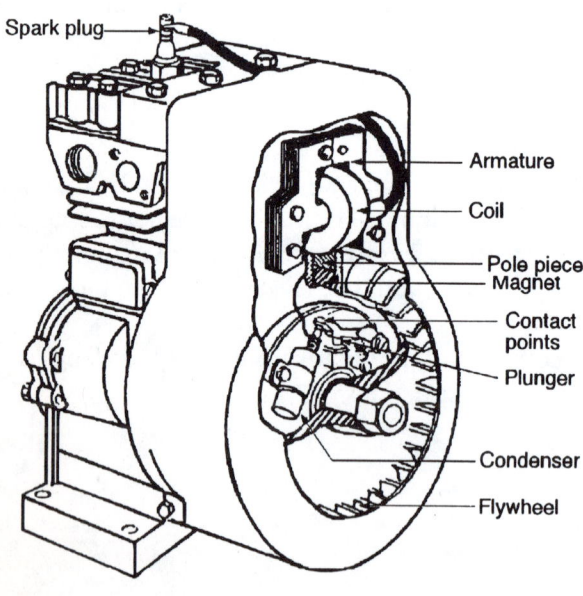

The principle components of the flywheel magneto ignition system

The magneto is designed to use the magnetic field of force that exists around a permanent magnet. This force field is used to generate electricity in a coil of wire. The most commonly used magneto on a small engine has a magnet located in the rotating flywheel and a coil attached to the block.

Producing Electricity Using a Magnet, a Conductor and Movement

It was discovered many years ago that you could produce electricity by relative movement between a magnetic field and a conductor. With the small engine magneto the conducting coil is usually stationary and the magnetic field moves.

If the conductor forms a complete circuit and the magnet is moved to cut across the conductor at a right angle, the lines of magnetic force will induce a voltage in the conductor. The flow of current will be very small, because the magnetic field only cuts one piece of wire.

To increase the induced voltage you can:
• increase the number of conductors cut
• increase the strength of the magnetic field
• increase the relative movement between the two (keeping the cut at 90°).

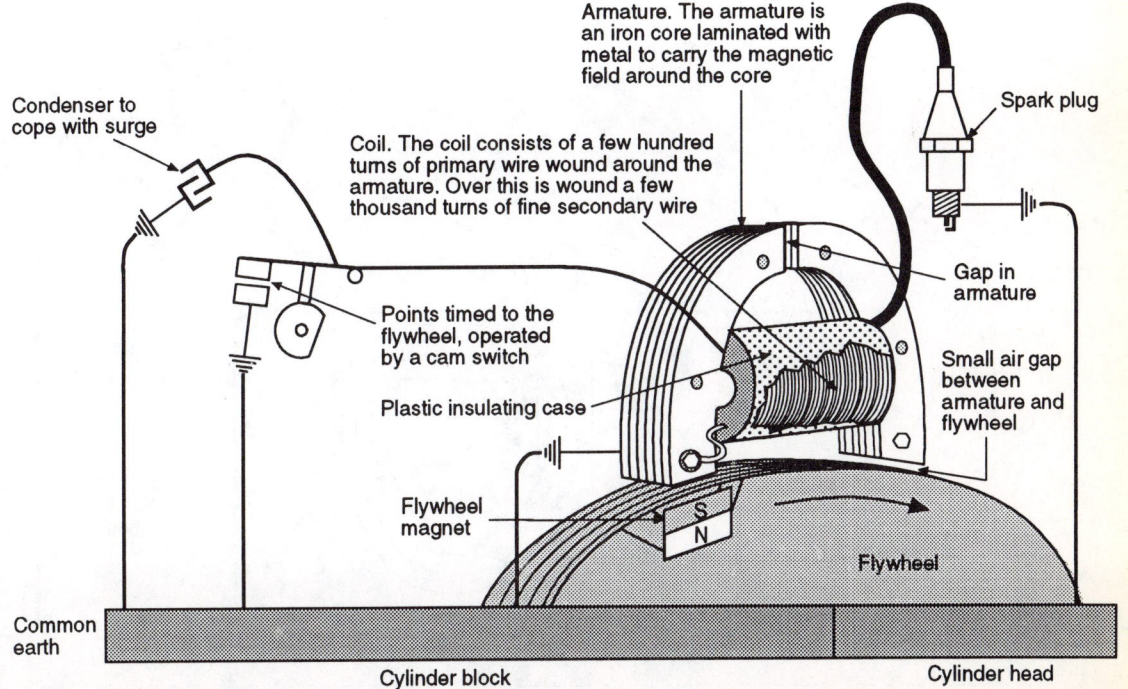

To increase the number of conductors cut you make a coil. The wires are coated with a varnish to provide insulation between them.

Voltage is only induced when the magnetic field is moving across the coil. If the magnetic field is moved slowly through the coil, the voltage will be low. If the magnetic field is moved quickly through the coil, the voltage will be high.

Magnetic lines of force travel more easily through iron than air, so the coil is wrapped around a laminated core of iron the magnetic field will flow from north to south pole through the iron. Laminated cores rather than solid cores are used in magnetos, to reduce the effects of eddy currents.

Just as a magnetic field cutting across a conductor can result in a current flow in that conductor, similarly a conductor that has a current flowing through it will cause a magnetic field to develop around it.

To generate a high voltage, magnetos use a winding with a few hundred turns of heavy wire to induce a voltage into a winding with a few thousand turns of very fine wire.

This is termed a step up transformer. For every one turn of the primary windings there are about 100 turns of the secondary winding, so that the step up ratio is 1:100.

Magnetic induction: Step one

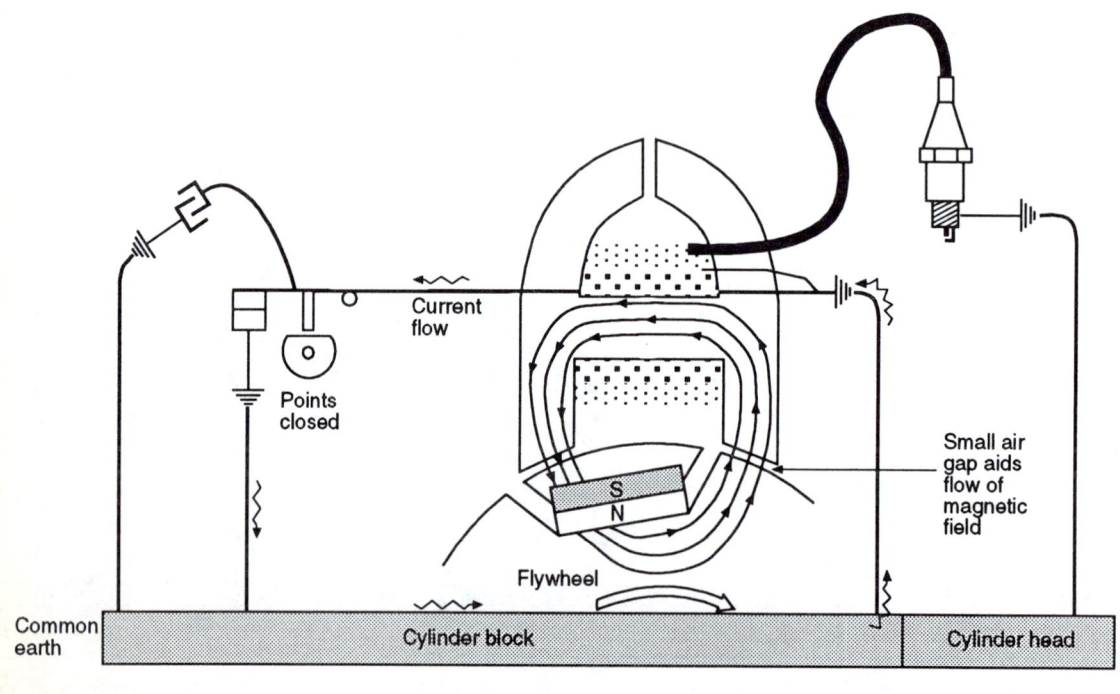

78

Step One

As the flywheel rotates, the magnetic field flows from the north pole to the south pole via the armature, cutting across the two coils. The magnetic field does not travel to the top of the armature and across because the air gap at the top offers too much resistance, so it takes the middle path.

The points are closed to complete the primary circuit, a voltage is induced and current flows in the circuit. This creates a magnetic field around the primary coil. The secondary circuit has the gap at the spark plug and the weak magnetic field from the magnet will not induce a strong enough voltage in the secondary circuit to jump the gap.

Step Two

As the flywheel rotates to this position the magnetic field of the flywheel magnet reverses its direction at the base of the armature legs. Because the primary winding carries a current, its magnetic field opposes the field from the flywheel magnet and forces it to find another path, which is over the top of the coil and across the air gap at the top of the armature.

But as the flywheel passes, the pressure is on the magnetic field of the primary windings to oppose the magnetic field of the flywheel magnet.

Magnetic induction: Step two

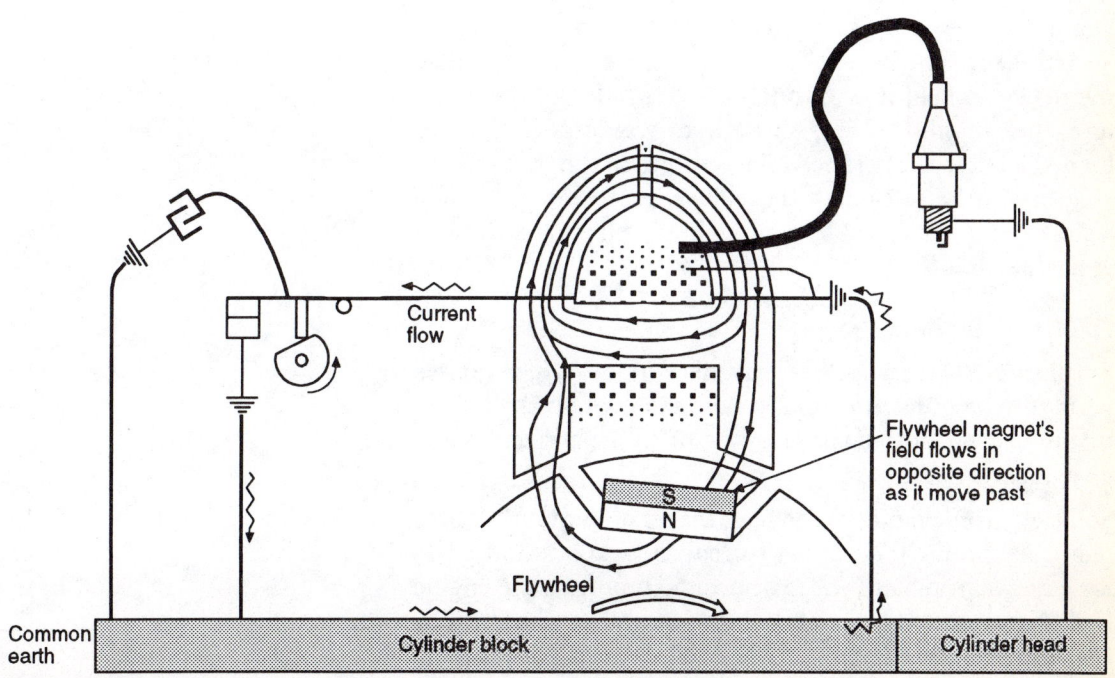

79

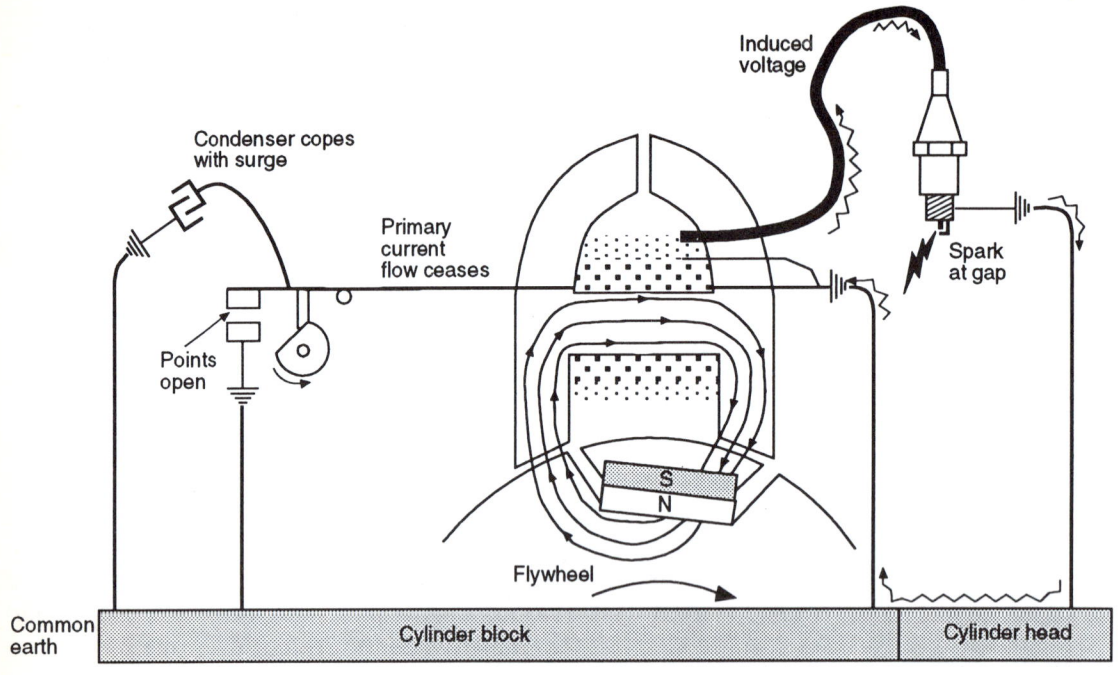

Magnetic induction: Step three

Step Three

With maximum pressure on the magnetic field of the primary circuit to oppose the magnetic field of the flywheel magnet, the points open. The current through the primary winding ceases and its magnetic field quickly collapses, allowing the magnetic field of the flywheel magnet to reverse the magnetic field flow through the centre leg of the armature, thus rapidly cutting across both windings inducing a high voltage in them. The voltage surges in the primary circuit, but in the secondary the voltage reaches over 10 000 Volts to jump the gap at the spark plug, producing a spark to ignite the mixture.

Electronic Ignition

Small engines today use an electronic ignition, which houses parts such as transistors, that do the same job as the points, that is, to switch off the primary circuit at the correct time to give a spark to the plug.

Some electronic modules can alter the ignition timing. These retard the timing for starting to reduce kickback and allow for easier starting and will advance the timing as the engine builds up speed. The module can be located at the coil or elsewhere on the engine.

An electronic module can also be fitted to most older engines with points by cutting the wire from the coil to the points and wiring it in.

Stop Switches

Engines are usually stopped by a simple switch on the primary circuit to earth or a switch which earths the high tension voltage at the spark plug.

Ignition Timing

On most four stroke small engines the ignition produces a spark at the plug each time the piston goes up. So on a four stroke, a spark is also produced on the exhaust stroke as it approaches T.D.C. This is referred to as a waste spark.

Most engines do not have adjustments to alter the time when the spark occurs, although some other engines may have adjustable plates that the points attach to. Timing on most small engines is fixed at about 25° before T.D.C., which is about when the piston is 3 mm from T.D.C. on the compression stroke.

Spark Plugs

The purpose of the spark plug is to provide a spark at the gap to ignite the petrol/air mixture near the end of the compression stroke to start the combustion process.

A spark plug consists of a metal centre electrode that sticks out of the engine on one end. The other end is found in the combustion chamber next to an earth electrode. The centre electrode is insulated from the head so that the high tension voltage from the coil can travel down the centre electrode and not escape, and then jump the gap across to the earth electrode making a spark to ignite the mixture.

Thread

The most common spark plug thread diameter is 14 mm with a 1.25 mm pitch. Very old engines used various threads, but metric threads have been used on spark plugs for many years.

Reach

Reach is the length of the thread that screws into the cylinder head. Plugs usually have a short, medium or long reach.

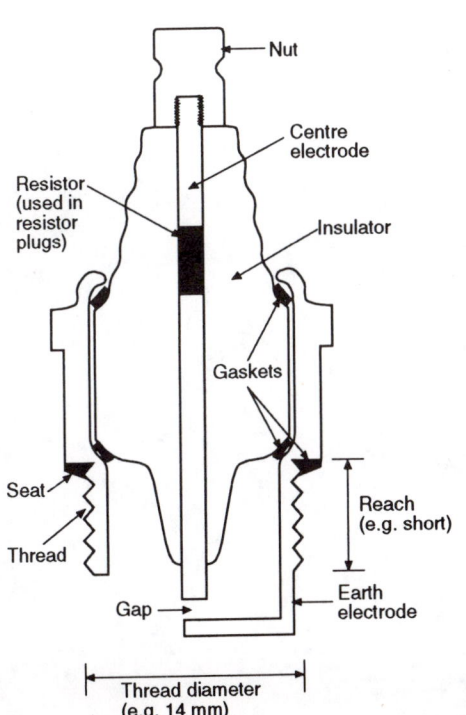

Spark plug features

If the correct reach plug is not used then serious problems can occur.

If a plug with a long reach is screwed into a head designed for a short reach plug, mechanical damage could occur, or carbon build up on the threads that extend into the combustion chamber could make it difficult to remove.

Resistor and Non-Resistor Plugs

Most vehicles today are fitted with resistor plugs. The resistor is fitted into the centre electrode. Many small engines still use non-resistor plugs.

Heat Range

The centre electrode at the plug gap has to operate at a temperature of between 400 and 800°C to function correctly. Too cold and it would foul up, too hot and it would cause pre-ignition.

To control this temperature the plugs have different length insulator tips to vary the length the heat has to travel from the tip to the cylinder head.

Seat Sealing

Where the plug seats against the head the seal is either:
• a gasket type or
• uses a tapered seat.

The two types are not designed to be interchangeable.

Ignition System Maintenance and Repair

As with any electrical system, it is very important to keep the ignition system dry and clean so as to reduce the problems that could occur.

All connections must be kept clean and tight so that the circuits have minimum resistance and the current flows freely. One dirty or loose connection can cause a problem.

The ignition system requires most maintenance at the spark plug where it is working in the combustion chamber. Engines with electronic modules instead of breaker points usually give little trouble. Breaker points are a problem because they wear and are burnt by the arcing as they open. Condensers absorb most of the arcing at the points, but not all.

Spark Plugs

Spark plugs should be cleaned every 50 to 100 hours of operation and replaced once a year at the start of each season.

Some manufacturers recommend that their spark plugs not be sand blasted nor the electrodes filed because a protective coating is applied to the electrodes and of the risk of sand lodging in the plug and dropping into the engine.

Normal cleaning entails cleaning off carbon deposits and resetting the gap, which is usually between 0.7 to 0.8 mm.

If the electrodes are badly fouled or burnt and you are not in a position to obtain a new spark plug immediately, you may have to use a file or emery on the electrodes to get yourself out of trouble.

A fouled plug may give a spark outside the engine, but inside it may fail.

If the electrodes are still square rub the surfaces where the spark jumps to clean away the fouling until clean metal is seen.

If the electrodes are burnt and rounded they should be filed flat. Electricity prefers to jump a gap that has clean square edges, rather than one that has round fouled edges.

Use a points file or a thin warding file. Bend the earth electrode a minimum amount to reduce the chances of it breaking off. Check the gap after cleaning.

Always carry spare new spark plugs to fit the engine you use, because they are the weak link in the ignition system and quite often a new plug is all it needs to fix it.

Always use the spark plug for your engine that is recommended by the engine manufacturer.

Breaker Points

Checking, adjusting, cleaning or replacing the points usually entails removal of the flywheel, for they are usually positioned behind it. When the points are a problem on older engines some people then fit an electronic module which usually fits to the outside of the engine.

To remove the flywheel firstly remove the parts that cover or surround it such as the starter assembly and shroud.

Then remove the nut that retains the flywheel; normally it is a right hand thread, but a few use a left hand thread. Generally speaking if the flywheel is used to spin the engine to start, the thread is right handed.

0.76 mm feeler guage

Setting the spark plug gap

(Please note that a spark plug analysis chart is printed on the inside back cover)

To correctly remove the flywheel a puller should be used. If a puller is not used and a hammer and levers are used there is a risk of damaging the engine.

Flywheels are either on a tapered or parallel shaft and located with a key.

Never hit the flywheel with a hammer because it may shatter when the engine is working.

If unable to dislodge the flywheel you may have to take it to a dealer or workshop.

Behind the flywheel, usually under a cover, are the points and condenser.

The points are mechanical switches and therefore they burn and get dirty and the rubbing block wears down, so they can be a problem.

The contact points are usually made from tungsten.

They can be filed and cleaned using a points file and fine wet and dry emery, or replaced. The points may come as a single unit or may come as two pieces with a contact point as part of the condenser.

When removing the points note the position of insulating washers and the wires.

Fit the new points in the reverse manner to the removal.

Adjusting the Points

A well fitting screwdriver and the right thickness feeler gauge are needed. Point gap is usually about 0.5 mm.
- Turn the engine over until the points are fully opened.
- Loosen the adjusting screw(s).
- Place the feeler gauge between the contact points.
- Close the contact points onto the feeler gauge.
- Tighten the adjusting screw(s) then feel the drag of the feeler gauge in the gap.
- The drag should only be slight, if not, readjust the gap.
- When gapped correctly, clean the points with a clean rag.

Checking Ignition Timing

Before fitting the flywheel the ignition timing can be checked on some engines. To check the timing you can use marks on the engine or if not applicable, note the opening of the points in reference to the position of the piston on the compression stroke.

When the piston is approximately 3 mm from T.D.C. on the compression stroke, the points just open to produce the spark.

To check the piston position you can gauge it through the spark plug hole or in some cases the head may have to be removed.

Refitting Flywheel

Before refitting the flywheel, check the condition of the key and keyways. Some engines use a key made from aluminium alloys. Do not replace it with a steel key.

Clean the magnet area with emery to remove rust and dirt.

Armature Air Gap

With the flywheel fitted and tight, check the air gap between the magnet and the armature.

The gap should be from 0.2 mm to 0.3 mm.

If it needs adjusting the armature can be moved by loosening off the retaining bolts. Be careful with these bolts for they are usually very small.

With the armature loose place a non-magnetic shim of about 0.2 mm to 0.3 mm between the armature and the magnet and tighten the bolts. With the air gap adjusted, turn the flywheel to ensure it does not strike the armature.

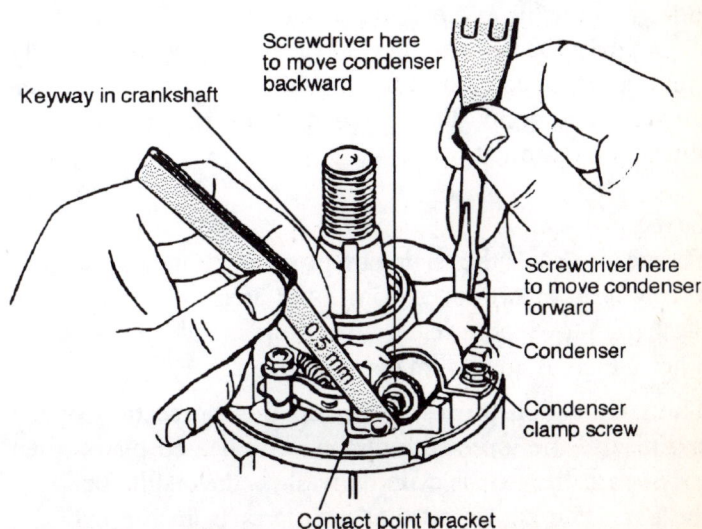

Keyway in crankshaft

Screwdriver here to move condenser backward

Screwdriver here to move condenser forward

Condenser

Condenser clamp screw

Contact point bracket

Setting points gap (see page 84)

Chapter 7

Lubrication System

Lubrication oil has four main purposes: it lubricates, cools, cleans and seals. It also cushions loads, reduces power losses and protects against rusting.

In the four stroke engine the oil is in the crankcase, which is similar to the sump of a vehicle engine. In a two stroke engine the oil is generally mixed with the petrol.

Lubrication

Whenever one surface moves over another they cause friction, which results in heat and wear. The main job of oil in the engine is to reduce friction. The oil provides a thin film that separates the moving metal surfaces and keeps the contact of metal against metal to a minimum.

If the engine runs with a low supply of oil, the heat of friction builds up rapidly. Parts like the piston sliding against the cylinder, and the big end bearing surface of the connecting rod sliding over the big end journal on the crankshaft, rapidly build up heat and start to fail, metal touches metal and seizure can then occur. The two areas just mentioned are usually the first affected.

Cooling

The oil also cools the engine by providing a means of transferring heat. Heat moves from a hot area to a colder area. Heat will move more readily from one metal surface to another if there is an oil film between them.

The combustion process heats up the top section of the engine and the top of the piston. The heated piston then transfers its heat to the cylinder wall via the oil film between the two. The cooler oil that circulates from the cylinder wall back into the oil supply in the crankcase also carries heat away from the piston. The heat of the oil in the crankcase is also transferred to the cooler metal of the crankcase.

Cleaning

The oil as it circulates through the engine tends to wash off and carry away carbon, dirt, pieces of metal and other foreign particles and deposits them in the crankcase.

Larger particles will stay at the bottom and some will be trapped in a screen or oil filter if one is used. Lighter particles will tend to circulate with the oil. Water also mixes with the oil. Water is a by-product of the combustion process and usually goes out the exhaust as a vapour. For every litre of petrol burned, about one litre of water is generated in the combustion process and some of this water finds its way into the oil. Water can also result from condensation when operating a cold engine. Water and dirt can also find their way in through the ventilation system. Water mixing with the oil forms a sludge.

Adding detergent to the oil can help in the cleaning process.

Sealing

Oil also provides a seal between the rings, piston and cylinder to help seal in the pressures that are applied to these areas, such as on the intake stroke when a good seal is necessary to draw in the mixture.

Cushioning Loads

Oil acts as a shock absorber. It cushions loads such as the load on the connecting rod from the stop/start action of the piston and the loads that are applied from the combustion process on the bearing surfaces.

Reducing Power Losses

As oil reduces friction it also reduces power losses. The lower the level of friction the less the power consumed internally to run the engine.

Protecting Against Rusting

Oil covers the metal parts and reduces the rusting effect.

Oil Viscosity and Service Rating

Engine oils are mainly products of crude oil. Some oils used now are synthetic.

Oil viscosity refers to the ability of the oil to flow. An oil with a low viscosity flows easily and an oil with a high viscosity is thicker and flows slower.

Oil viscosity is rated by the Society of Automotive Engineers (SAE) for both winter and summer use. Winter grades are SAE5W, 10W and 20W. Summer grades are SAE20, 30, 40 and 50. The higher the number, the higher the viscosity and the thicker the oil.

Most engine oils used in Australia are multigrade so that the same oil can be used all year round. For example SAE20W-40 has the same viscosity as SAE20W when it is cold and the same viscosity as SAE40 when it is hot.

The engine manufacturers usually specify the viscosity and service ratings for the oils that give the best performance and longest life for their engines.

An oil service rating is a set of letters (usually two) that are printed on an oil container to indicate how well that oil will perform under certain operating conditions. It is a performance standard that is set by the American Petroleum Institute (API).

The ratings for petrol engine oils start at SA and go up; they currently go up to SG. For example, a high quality detergent oil would be classified for service SC, SD, SE, SF, SG. Oil service ratings for diesel engines start at CA.

Some engine oils today are multipurpose and the one type may be suitable to use in a range of engines from the small engine up to the diesel tractor. Always use the oil type recommended by the manufacturer.

Oil Seals

Oil seals are used where the crankshaft comes out of the crankcase. The oil seal prevents the oil from leaving the crankcase and prevents foreign matter entering the crankcase. The lip of the seal must face the oil.

Gaskets

A gasket is a piece of material placed between two parts to prevent leaks. When the parts are tightened together any irregularities will be taken up by the gasket material to produce a leakproof joint. Most of the gasket material used on small engines is special oil resistant gasket paper of various thicknesses.

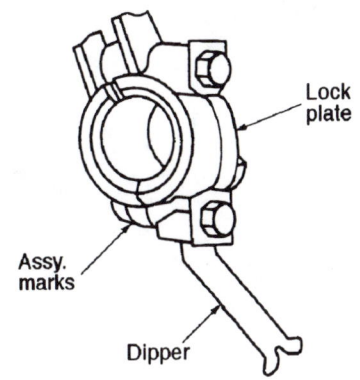

Oil dippers splash lubricant around crankcase

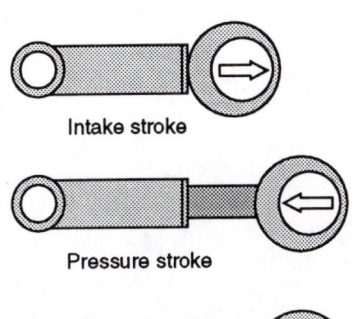

A barrel type oil pump connected to the camshaft

Methods of Lubrication in Four Stroke Engines

Four stroke engines use a wet sump, where the crankcase holds a supply of oil.

Several different methods are used to get the oil to the moving parts of the engine. The most common method is to splash the oil around, so that the parts get a good drenching of oil. Some engines use a simple oil pump to pressure feed some areas and more modern engines use a full pressure system with an oil filter.

Splash Method

This is the simplest and most common method. Splashing is produced by a dipper on the cap of the connecting rod. As it strikes the oil every revolution, the oil is splashed around the inside of the engine. Dippers are used on horizontal crankshaft engines.

In engines with splash systems the plain bearings that are used on the connecting rod small and big end and the crankshaft mains usually have oil holes to allow entry for oil.

Engines with vertical crankshafts, such as those used on lawn mowers can use a rotating slinger, driven by the camshaft, to splash oil around.

Pressure Method

Some engines use pressure differences in the crankcase to pump oil around. Others may use a barrel type pump which is operated by a special lobe on the camshaft. Others have small gear or rotor pumps that pick up the oil through a screen filter to pump the oil around.

Modern engines are using full pressure lubrication with a spin on oil filter similar to a vehicle engine's.

Crankcase Breather

A single cylinder engine, with only one piston travelling up and down the cylinder, can have high and low pressure surges in the crankcase. Blow-by past the rings also adds to crankcase pressure. Because of this the crankcase must have a means of breathing, to prevent pressure build up and oil losses at the seals and gaskets. A crankcase breather maintains a low pressure in the crankcase and prevents a pressure build up.

Most engines now have a crankcase breather pipe running from the engine, normally at the tappet cover, to the base of the air cleaner assembly. On older engines the breather pipe was vented to the outside of the engine, but this method can allow entry of dirt into the engines and atmospheric pollution. By venting to the entrances to the carburettor the crankcase fumes are burnt in the engine and dirt is prevented from entering the engine.

The breather has a disc valve that allows the pressure to leave the crankcase when the piston travels down the cylinder, but blocks any air being drawn back into the crankcase when the piston goes back up the cylinder.

Lubrication System Maintenance

The engine manufacturer's recommendations for maintenance should always be followed.

Checking Oil Level

The oil level is usually checked with a dipstick as on a vehicle engine or with a level plug.

Some manufacturers also fit devices that will cause the engine to shut down to prevent damage if there is insufficient oil in the engine.

- Oil level should be checked every day or after every five hours of engine use.
- The engine should be level for an accurate reading.
- Small engines only hold from about 500 mL to around 2 L of oil in the crankcase, so it is critical that the level be checked on a regular basis as recommended. Engines are designed to use up some oil.
- It is good practice to check the oil level when the petrol tank is filled.
- If the engine is fitted with gear reduction, the gear reduction oil level should also be checked as recommended.

Changing the Engine Oil

Oil should be changed according to the manufacturer's recommendation which is usually based on hours of use or time elapsed. Under normal operating conditions it is every 50 hours but sometimes up to 100 hours. Under abnormal operating conditions such as heavy loads, dusty work or high temperatures, the hours recommended are usually halved.

The oil should be changed at least once a year if the hours do not build up. At the end of the working season the oil is contaminated and should be drained while the engine is hot.

During the off season, condensation usually contaminates the oil. It is a good idea to run the engine when it comes out of storage to warm it up, then drain the oil out and replace it prior to starting the season.

* Do not be stingy with oil changes because good clean oil is cheap insurance. (How much does one litre of oil cost?)
* Refill the crankcase to the required level with oil as recommended by the engine manufacturer.
* Use clean oil handling equipment.

Changing Gear Reduction Oil

* Oil should be changed as recommended by the manufacturer. Usually every 100 hours of use or once a season.
* Some models require the housing to be loosened or the use of a drain plug.
* Refill to the level plug with the recommended type of oil, which in most cases is engine oil.

Changing Oil Filters and Cleaning Screens

Late model engines that use spin on oil filters have them replaced in a similar manner to those fitted to vehicle engines.

1. Unscrew the filter (anti-clockwise) with a filter removing tool.
2. Clean the gasket mounting face on the engine.
3. Follow fitting instructions or those below:
   * Smear the gasket with fresh oil.
   * Turn the filter until the gasket just contacts the mounting face.
   * Tighten it two thirds of a turn.
   * Check for leaks on start up.

Cleaning an oil filter screen may require the removal of a part to gain access. Clean the screen and refit it to engine. Screens are usually located in the bottom of the crankcase.

Mixing Oil with Petrol for Two Stroke Engines

Instead of having an oil reservoir, most two stroke engines are lubricated by oil mixed with the fuel. The mix ratio for two stroke fuel must be appropriate for the engine. Follow the manufacturer's instructions.

- Air cooled two stroke engines run hotter than water cooled two stroke outboard boat engines so you must make sure the oil you use is the correct type.

- Previously, ordinary engine oil was mixed with petrol for two stroke engines, but now special oils have been developed.

- It is important that petrol containers that have the two stroke mix in them are clearly marked so you do not accidentally put straight petrol into your two stroke engine and cause engine damage due to lack of lubrication.

- There are different rates of mixing for petroleum based and synthetic oils.

Mix as per the engine manufacturer's recommendations. The following is a general mix ratio to suit most two stroke engines.

Two stroke air cooled petroleum based oil should be mixed at 20 parts petrol: 1 part oil or 1 L: 50 mL. Two stroke air cooled synthetic oil should be mixed at 40 parts petrol: 1 part oil or 1 L: 25 mL.

Thoroughly shake the container when mixing and before use.

Chapter 8

Cooling System

The purpose of the cooling system is to ensure that the engine does not exceed its normal operating temperature.

Two types of systems are used to cool internal combustion engines: air and liquid. Vehicle engines are normally liquid (water) cooled and small petrol engines are normally air cooled.

Small petrol engines use a forced draught system where a fan forces air to flow over the cooling fins that are part of the cylinder and the cylinder head to carry away the excess heat from the engine.

Operating Principles

When petrol is burnt in the combustion chamber, heat results. Part of this heat, about one quarter to one third is used to drive the piston down. The rest is lost. About one third goes out the exhaust. But about one third is lost to the engine and must be dissipated by the cooling system.

For an air cooling system to work effectively it must have fins. Fins are thin pieces of metal that are cast as part of the cylinder and the cylinder head. These ensure that a large surface area is exposed to the air, and sufficient heat from the metal can be transferred into the cooler air that flows over it.

To force the air over the cooling fins the flywheel has blades that cause the air to move when the flywheel rotates. The air is drawn in through the blower housing screen at the flywheel centre and the rotating flywheel acts as a fan and forces the air over the cooling fins and out the opposite side of the engine. The blower housing shrouds, baffles and deflectors guide the air flow over the fins.

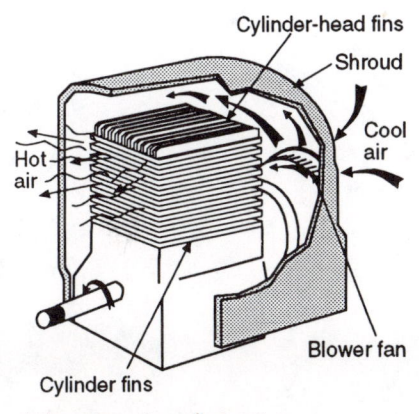

Action of an air cooling system

Maintenance

The cooling system on many small engines is sadly neglected.

It is important to clean the systems at least once a year or more often if the engine is operating in dusty or chaffy conditions. If the system is neglected the engine can overheat and fail with drastic results.

If the engine leaks oil, dust can accumulate around the fins and cause a hot spot. This can lead to a partial seizure. If the engine leaks oil, fix it quick!

To clean the fins, the blower housing and baffles etc. may have to be removed. Avoid the use of water. Use a stick if possible to avoid scratching the metal fins, because a scratch encourages dirt to stick. Use compressed air to get in between the fins. Insects and mice like to build nests in the fins and associated areas during the off season. Nests and dry vegetable debris are also a fire hazard.

Chapter 9

Exhaust System

The purpose of the exhaust system is twofold: firstly to provide a passage for the burnt gases to escape to the atmosphere once they leave the engine cylinder, via the exhaust valve on a four stoke engine or exhaust ports on a two stroke engine; secondly to provide a means of reducing the loud noise that the engine produces, as a result of the combustion process, to an acceptable level.

Mufflers

Mufflers are devices used to muffle sound. They absorb the noise generated by the engine and the expanding gases and delay the hot gases slightly so that they hit the cooler outside air in a steady stream instead of immediately. This has a tendency to reduce the loud banging noise.

Mufflers are designed to present a minimum of back pressure to the engine. Excessive back pressure can effect performance.

They are attached to the cylinder block or head.

Muffler design is usually of the baffle type or tube type. The baffle type allows the burnt gases to expand and cool before moving into the atmosphere. The tubes inside the tube type have holes to allow for the expansion of the gases.

Because the exhaust gases are ejected from the engine in surges, the burnt gases spend a split second in the muffler before being ejected by the next surge of burnt gases from the engine.

Two stroke engines are louder than four stroke engines so they require special mufflers.

Some spark arrester type mufflers are fitted to engines such as those on firefighting units and chainsaws.

Maintenance

Exhaust systems require cleaning out on occasions, more so on two stroke engines. Carbon builds up around the port areas and chokes the engine so that the burnt gases can not flow out correctly. As a result the back pressure increases, an adequate fresh mixture can not flow into the cylinder and the engine loses power.

To clean the ports, remove the muffler and turn the crankshaft so that the piston skirt covers the ports. Use a stick to avoid scratching the piston and dig out the carbon.

The bolts that hold the muffler on tend to vibrate loose so keep a check on their tightness.

Mufflers are subject to corrosion and rust like those on vehicles and should be checked for this also.

If fitting a screw in muffler, place some type of anti-rust product on the thread to assist in easy removal of the muffler in the future. Clean the spark arrester (if fitted) when recommended.

Chapter 10

Starting System

The purpose of the starting system is to rotate the crankshaft to operate the ignition and start the engine.

Mechanical starters, usually rope rewind starters, are still the main method of starting small petrol engines. Some of the slightly bigger petrol engines use electric starters that are similar to the systems in vehicles.

The rope rewind starter uses a light spring to wind the rope back into the starter housing after it has been pulled out to start the engine. It also has a ratchet device so the engine flywheel does not turn the starter when the engine is running.

The starters are built to last the life of the engine, but if the engine is hard to start, the starter suffers and the rope may break or the spring fail.

We shall look at how to dismantle and reassemble two typical rewind starters that are fitted to horizontal crankshaft engines. The first is the type used on many Briggs and Stratton engines and the second is the type used on the Honda GX160K1 engine and other similar models.

Briggs and Stratton Starters

Safety note: Wear eye protection and gloves when handling a rewind spring.

There are two easily made tools that are used. One is a rewind starter tool for use with the pulley. This is a metal bar that will fit inside the pulley to turn it and coil or loosen the springs. It has two holes to insert two bars – one for you to turn, the other to lock the bar and pulley in the starter housing.

The other tool is called the rope inserter and can be made from a piece of music wire or spring wire.

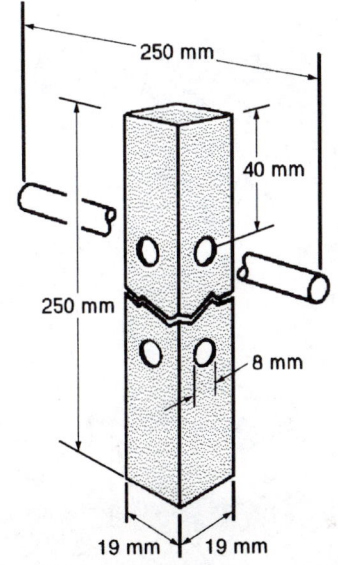

Starter rewind tool

First remove the blower housing from the engine by undoing the retaining bolts. To work on the starter assembly in the blower housing, it is a good idea to secure the housing to a bench with a G clamp or use a large vice (with rag material in the vice jaws).

When the rope is pulled out, a ratchet device locks to rotate the crankshaft. When the engine starts, the spinning crankshaft unlocks the ratchet device so that the crankshaft does not rotate the starter pulley. The ratchet device drives in one direction and disengages in the opposite direction.

Replacing the Rope

1. If the rope has broken, pull the remains of rope out of the pulley.

2. If the rope is intact, pull the handle so the rope is pulled all the way out and the pulley has wound up the rewind spring tight. Then insert the rewind starter tool into the centre of the pulley hub and slide a bar through the lower hole to lock against the housing stays. Let the rope partly rewind.

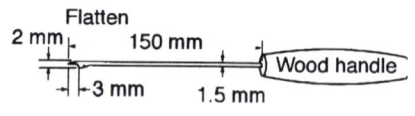

Rope inserter

3. Undo or cut the knot at the pulley end and pull the rope out. Remove the handle from the rope.

4. Check if the spring is fully wound up by inserting another bar in the top hole. Take the load and remove the bottom bar, then turn the tool anti-clockwise until the spring is fully tight, then back off one turn or until the hole in the pulley for the rope lines up with the rope eyelet in the housing. Now insert the lower bar back where it was, to hold the pulley from rotating.

5. Cut the new rope to the required length. (This can be measured off the old length or you can consult the local dealer.) Rope is usually about 120 cm long. Burn the ends with a match. Wipe it with a clean dry rag, using caution, to reduce the end to a usable point and prevent swelling and unravelling.

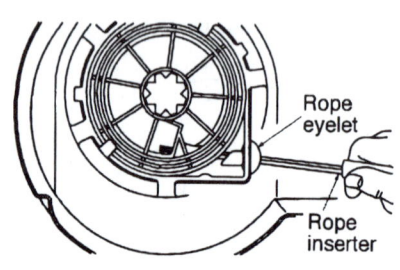

6. Using the rope inserter, push the rope through the rope eyelet in the housing then up through the hole in the pulley. Tie a simple knot in the end as shown.

    If using an old style pulley with a guide lug, make sure the knot does not contact the bumper tags.

    If using the current style make sure the knot is pulled down into the knot cavity.

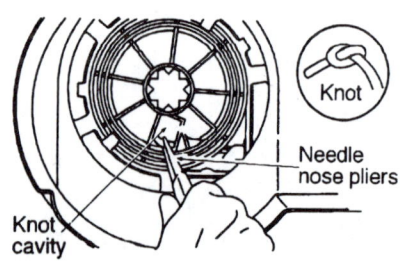

7. Fit the handle to the other end of the rope and pull the pin and knot tightly into the handle.

8. Take the pressure on the rewind starter tool and remove the lower bar. Slowly release the pulley so the rope fully rewinds.

Test the rope to see that it pulls out freely.

Refit the starter to engine ensuring that the starter ratchet freely engages the hole in the pulley.

Mending or Replacing the Spring

1. With the housing secured, repeat steps 2 and 3 above.

2. Using the rewind starter tool slowly release the tension of the rewind spring. With the spring relaxed remove the tool from the pulley.

3. Grasp the end of the spring at the outer edge of the housing with a pair of pliers and pull the spring out as far as it will go. Then turn it 90° and unhook it from the pulley. The end of the spring will come out of the housing.

   The spring can also be unhooked from the pulley when the pulley is removed from the housing.

4. Remove the pulley from the housing. You may have to bend the holding tangs if present.

5. Using gloves straighten the spring by bending it with your hands to provide more tension and to allow easier installation. Inspect the spring if it is to be re-used or obtain a new spring.

6. Clean the rewind housing, pulley and rewind spring in a cleaning solvent. Dry with a clean rag. Oil the spring and insert either end of the spring through the slot in the blower housing and hook it into the pulley. Place a dab of grease on the spring side of the pulley.

7. Replace the pulley into the housing (bending the tangs back). Check that the pulley has sufficient clearance.

8. Place the rewind starter tool in the pulley hub and using the top bar wind the pulley anti-clockwise, feeding the spring through the slot until the end of the spring is locked into the housing. Continue to wind the pulley until it is tight, then back the pulley off one turn or until the hole in the pulley for the rope and the rope eyelet in the housing are aligned.

9. With the spring wound up in this position place the other bar in the lower hole to lock the pulley so it stays in this position.

10. Fit the rope following steps 6, 7 and 8 from Replacing the Rope.

Honda Starters

> Safety Note: Wear eye protection and gloves when handling the recoil spring and associated parts.

1. Remove the recoil starter assembly from the fan housing.

2. Pull the knot out of the starter grip and untie it. Slowly let the starter reel unwind to release the energy of the spring. Use gloves for safety.

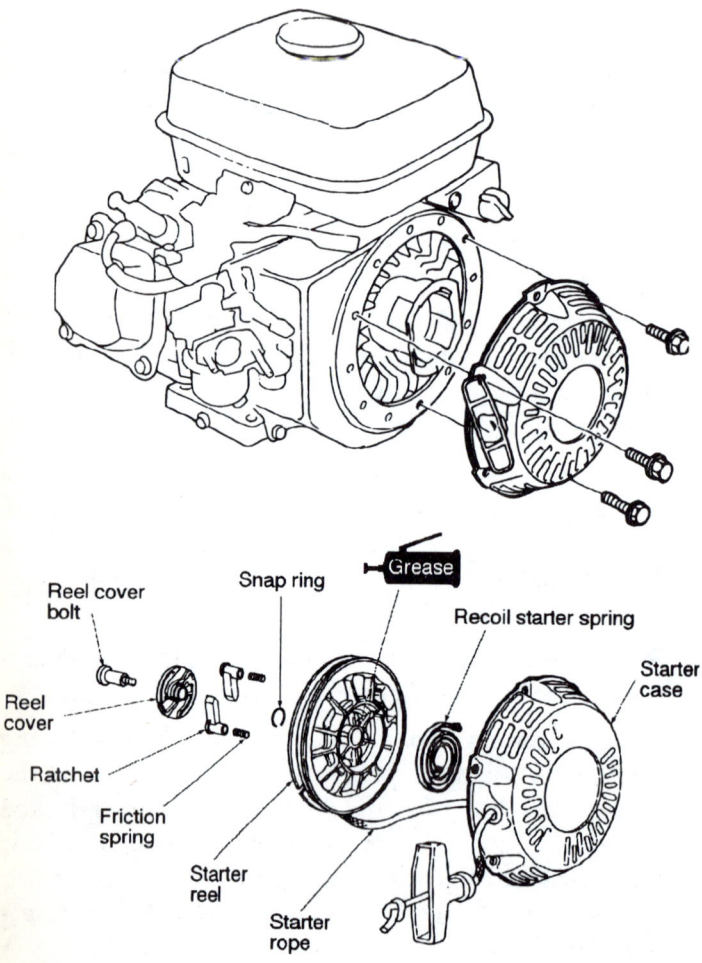

3. Remove the reel cover bolt and carefully remove the parts and lay them out in the order of removal.

   Be careful if removing the recoil spring.

4. Clean all parts.
   Check all parts for serviceability.
   Check the ratchet for chipping and wear.
   Check the starter rope for fraying and wear.
   Check the condition of the recoil spring. It if has to be replaced handle it carefully with gloved hands.

5. Commence reassembly by inserting the hook on the outer side of the recoil spring into the groove inside the starter reel.

6. Pass the starter rope through the starter reel and tie it as shown. Wind the starter rope around the starter reel in the direction of the arrow. Leave approximately 30 cm of the starter rope outside the starter reel.

7. Install the starter reel on the starter case so that the spring inner hook is hooked to the case tab.

8. Hold the starter case and rotate the starter reel two revolutions in the direction of the arrow for preliminary winding.

9. Pass the starter rope end through the starter case rope guide and pull it outwards. Pass the starter rope through the starter grip and tie the rope. Do not separate the starter reel from the starter case, otherwise the return spring inside the case will come off, which could cause injuries.

10. Install the ratchet with the spring and reel cover. Tighten the reel cover bolt.

11. Check the operation of the ratchet by pulling the starter rope several times.

12. Ensure there is no dirt or debris where the starter assembly fits onto the fan housing. Refit the starter assembly to the fan housing in the best location for the starter grip.

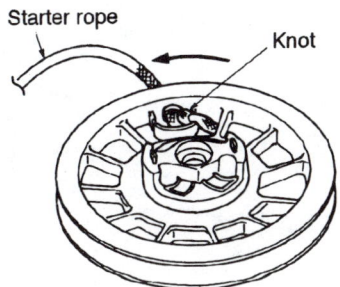

Starter rope

Knot

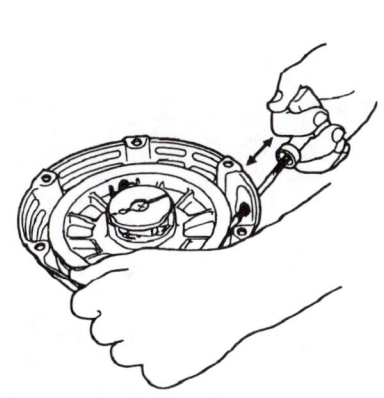

Chapter 11

Routine Maintenance

Routine maintenance is the regular care and attention that an engine needs so that it can perform its role in a safe and efficient manner during its working life.

Maintenance is not about fixing an engine after it breaks down, it is following the engine manufacturer's recommendations and looking after the engine to prevent it from breaking down and wearing out before its time. An engine is built to last only so long doing a certain job, but you have to maintain it correctly to get the maximum economical operating life.

Slack maintenance is a sure way of reducing the life span of an engine. A poorly maintained engine can cost you money if it fails during a busy season.

Engine manufacturers supply an owner's manual with each engine that they sell, to inform the owner how to operate and maintain the engine properly.

If an engine is resold the owner's manual should go with the engine to the new owner.

The main thing to note in any maintenance manual is the maintenance schedule, which lists the tasks to be performed and the service intervals in hours of operation or time elapsed. Stick to the maintenance schedule and you will get a good run out of your engine.

If there is a problem with a engine that is beyond your mechanical ability to rectify, take the engine to the authorised engine dealer.

General engine maintenance is explained in earlier sections of this book. Removing combusion deposits is an additional routine maintenance task.

Removing Combustion Deposits

Some engine manufacturers recommend that the cylinder head be removed every 100 to 300 hours of operation and the carbon build up (combustion deposits) be removed from the cylinder head, cylinder, top of the piston and around the valves. An excessive amount of combustion deposit will result in a loss of power, short valve life and the possibility of the piston and valves hitting the build up and causing mechanical damage. If you are mechanically proficient with suitable tools and a workshop manual, you can do the job yourself; otherwise take it to an authorised dealer.

The following are the basic steps to follow to remove the combustion deposits on a typical side valve engine:

1. Thoroughly clean down the main body of the engine and safely disable the ignition and fuel systems.

2. Remove the blower housing, air shroud etc. to enable the head to be removed.

3. Keep all head bolts in the correct order, because different length bolts are often used. An easy way is to use a piece of cardboard and draw the shape of the head on it and punch holes to fit the bolts in as you remove them.

4. With the head removed you will note the carbon build up. Before attempting to clean it off, take the piston to the top of the compression stroke by slowly rotating the crankshaft in the direction of rotation. Then take the piston about 10 mm down on the power stroke. (Both valves are fully closed, due to the decompression devices not being engaged.) Get some normal chassis grease and smear it between the piston and the cylinder wall to prevent carbon dropping down between the piston and the cylinder. (This can result in a scratched bore.) Do not rotate the crankshaft now until all the deposits have been cleaned off.

5. Using a suitable scraper, such as a piece of hard plastic, remove the combustion deposits. Do not scratch the gasket surfaces. Do not dig into the aluminium surfaces or damage the top of the piston.

6. When the deposits have been removed the crankshaft can be rotated to move the piston down the cylinder slightly so that the grease can then be wiped off. The valves can then be fully opened to enable the valve faces and seats to be wiped clean.

7. Once the head is clean, the flatness of the gasket contact surface can be checked. One method is to use a machined

surface plate and a 0.1 mm feeler gauge. With the head held firmly on the flat surface of the plate, see if the feeler gauge will slip in between the gasket contact surface and the surface of the plate at all positions.

If the feeler gauge does not slip in at any point, the head can be considered satisfactory. Also look over the head for cracks and the condition of the thread for the spark plug.

If the feeler gauge does slip in at some point, the head needs attention to get it flat. One way is to use a sheet of about 320 grit wet and dry emery on the surface plate, pour on some kerosene, hold the head firmly on the emery and rub the head in a figure eight pattern, rotating the position of the head every 30 seconds. Only remove enough metal so that the feeler does not slip into the gap, otherwise the head can be ruined (the opening valves can strike the head).

The head can then be placed on the mating surface of the cylinder block and the feeler gauge used to check if the block surface is flat. The head surface can also be checked using a straight edge and feeler gauges.

The head surface can also be reclaimed at a machine shop.

8. Clean the head bolts with a wire brush and hand screw the bolts into their respective holes. They must screw freely all the way into their hole. Also check the thread of the bolts and that the bolts are not cracked or stretched.

   The holes should be precleaned using compressed air. Check that the holes in the head for the bolts are clean. If the bolts are to screw into an aluminium block they should be lightly coated with a graphite grease.

9. With all the parts clean and checked and the cylinder bore lightly oiled, the head can then be refitted using a new head gasket. When you buy the head gasket obtain the head bolt tightening torque figures from the dealer. Do not use any sealants on the head gasket.

   Install the head bolts, with air shroud, etc. in place, and finger tighten.

   Identify the torque tightening pattern to be used and tighten the bolts to half the final figure first, then go over them again to tighten them to the final torque tightening figure.

   Service the spark plug or fit a new one.

   Double check that everything is in place and tight.

   The engine should now be ready to start.

Chapter 12

Operating the Engine

Owner's Manual

Before you operate any engine read the owner's manual. It covers the points that the engine manufacturers want you to know about operating their engines. Also read the manual of any equipment that the engine may operate so that you are familiar with it too.

Safety

Always consider the safety aspects of operating a engine. Again the owner's manual covers the main safety issues and see Chapter 2.

Pre-start Checks

*   Be familiar with all the controls on the engine and how they work.
*   Know the names of the external parts of the engine and their functions.
*   Check that all the parts are on the engine and that they are fitted correctly.
*   Check that the air cleaner is clean and correctly fitted.
*   Ensure that there is not a build up of dry vegetation around those hot parts of the engine that could be a fire risk or cause the engine to overheat.
*   Check the oil level and top up with the correct oil if required. To check the oil, the engine must be level. If using a dipstick, remove the dipstick and wipe it clean then re-dip to obtain the reading. Clean around the oil plug or dipstick before checking and use a clean oil handling container and funnel.
*   If there is a reduction gearbox on the engine check it. Also check oil levels on equipment the engine may operate.

- Check the fuel level in the tank. Top up with clean fresh petrol. Always leave an air gap in the tank to allow for expansion. Use a clean funnel fitted with a gauze strainer. Handle petrol with caution. If filling a two stroke engine with petrol ensure that correctly mixed two stroke fuel is being used.
- Check any powered equipment is in good working order and remove the load from the engine if possible, so that the engine is easier and safer to start.

Starting the Engine

- Turn the fuel tap on (if fitted).
- Set the choke on, if the engine is cold.
- Turn the ignition switch on.
- Set the throttle control to just above the idle position.
- Beware of kickback when starting the engine.

Most engines today use some type of compression release so that they are easier and safer to start than older engines. Some electronic ignitions also alter the spark timing for easier starting. Kickback is the tendency of the piston to be driven back down the cylinder before the engine turns over due to the spark starting the combustion process near the end of the compression stroke. As a result of the piston being driven backwards the crankshaft quickly changes direction. With kickback, the starter handle can be torn out of your hand. Also, if the flywheel is slightly loose and fitted with a soft key, kickback can shear the key and put the ignition timing out. Older big displacement engines can give a powerful kickback if not started correctly.

To reduce the kickback effect on engines fitted with compression release devices, you should pull the starter rope handle slowly, until you feel resistance (the piston travelling up the cylinder on the compression stroke). Then pull the handle rapidly to build up momentum for the next compression stroke, thus reducing the kickback effect and the engine should safely start.

To reduce the kickback effect on older engines that do not have compression release devices fitted, particularly the bigger displacement engines, it is suggested that you slowly turn the crankshaft over until the piston is at the top of or slightly after the compression stroke, before pulling the starter rope handle to build up the momentum.

- When the rewind starter rope is pulled out, let it carefully back into the starter housing. Do not release it so that it flies back in quickly.
- If using a hand wound rope starter be careful of anyone standing near you as you pull the rope. Do not wrap a starter rope around your fingers, because of the danger of kickback.
- When the engine starts set the choke at about halfway and keep the engine speed just above idle so that the engine can warm up.
- Warm the engine for a few minutes before you place a load on it.
- In a few minutes the engine should run without the choke.
- Do not over-use the choke because the excess petrol washes the lubrication off the cylinder wall, rings and piston. The oil also becomes diluted.

Running the Engine

- A four stroke engine must operate fairly level for the lubrication and fuel system to operate correctly.
- When the engine is warm the speed can be increased to the operating level and the load applied to the engine.
- Most small engines are not left unattended, but if they are they should have some protection system to shut them down if a malfunction occurs. Some now have devices fitted so that if the oil level drops too low, the ignition will be automatically turned off to stop the engine.
- When operating the engine you have to listen for unusual noises and ensure the engine does not overheat or run low on oil.
- When operating the engine do not overload it or run it too quickly. The governor controls the maximum speed of most small engines, so do not tamper with the governor.
- Do not operate an engine in a confined space.
- Use a muffler fitted with a spark arrester if necessary.

Stopping the Engine

- Before stopping an engine, take the load off it and reduce the speed to idle, if possible, for a few minutes to let the engine cool down.
- If you stop the engine when it is at speed above idle, unburnt petrol accumulates in the cylinder.
- Do not use the choke to stop the engine for the same reason.

- Use the approved method to stop the engine, which is usually either a switch to earth the primary wire, or a switch to the spark plug to earth the (high tension) secondary wire of the ignition. Another method is to turn off the fuel tap so the engine runs out of petrol.
- Allow a couple of minutes after stopping an engine before you refuel it.
- Never refuel an engine while it is running.

Chapter 13

Troubleshooting

If an engine has the right fuel mixture, good compression and a good spark at the right time, it should run smoothly. The fuel system provides the mixture of petrol and air, the mechanical system takes in and compresses the mixture and the ignition system provides the spark at the right time to ignite the mixture.

When an engine gives trouble, the first step is to determine which of these three systems:
- the mechanical system
- the ignition system
- the fuel system

has malfunctioned.

Troubleshooting is a logical, step-by-step process of locating engine trouble. When you troubleshoot you examine or test the engine to determine the possible cause of its trouble. To be good at troubleshooting you must have a good understanding of the basic operating principles of the engine.

The first step is to establish the symptom, then check out the most common causes for the symptom, and look for a solution. With a good understanding of engine operating principles and using troubleshooting charts and reference books, you can confidently locate and correct many problems.

If the engine is playing up and the cause is not obvious, you should perform a quick check for compression, spark and fuel, which can be carried out in a few minutes and will indicate where the cause of the trouble lies. Before checking these three areas do obvious checks first such as:
- the oil level in the crankcase
- that all parts of the engine are attached
- that all wires are connected

- that there is fresh petrol in tank
- that the fuel tap is on.

Also, always keep in mind as you troubleshoot that the trouble may be with the equipment that the engine operates rather than with the engine.

Compression

Carrying out a compression check only takes a few minutes and is a reasonable indicator of the condition of the engine's mechanical system.

Compression Checks

Many small petrol four stroke engines use some type of compression release so that the engine can be started more easily. If the engine is so equipped, the recommended method to check the compression is to remove the blower housing and spin the flywheel backwards (anti-clockwise). (If you spin the engine backwards the compression release device usually does not work. Switch the ignition off or earth the high tension leads at the spark plug to prevent accidental starting whilst checking to protect the ignition system.)

- If the engine has good compression, the flywheel will rebound sharply.
- If the engine has poor or no compression, the flywheel will give slight or no rebound.
- Engine manufacturers will recommend how to check for compression and the results to expect.

If the engine is not equipped with a compression release device (or if the compression release device can be disconnected), a compression tester gauge can be used to check the compression. This is called a dry test.

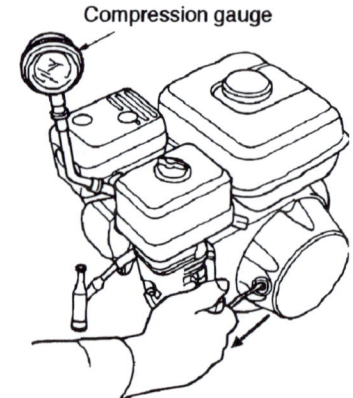

Compression gauge

- Disconnect the engine from the equipment it operates (if possible).
- Keep the engine at operating temperature (if possible).
- Clean around the spark plug and remove the plug.
- Set the choke fully off and set the throttle control to operating position so that the engine can breathe.
- Place the compression tester firmly in plug hole.
- Switch the ignition off or earth the high tension leads to protect the ignition system.
- Pull the starter rope so that the crankshaft rotates at about 600 rpm.

- Pull the starter rope several times until the gauge needle stops rising.
- Whilst cranking watch the needle on the gauge.
- If the compression is good the needle will climb quickly. If the compression is poor the needle will climb slowly to a lower peak point.
- Repeat the check several times. If in doubt use another reliable compression gauge.

A side valve engine should give a reading of about 410 kPa and an overhead valve engine should give a reading of about 620 kPa, if the compression is reasonable.

If the compression is lower than these readings you can carry out a "wet" test to check if the compression loss is in the area of the rings or the valves.

- With an oil can spout in the spark plug hole, squirt three pumps of engine oil onto the piston at the bottom of its cycle.
- Spin the crankshaft a few turns to spread the oil over the bore and rings.
- Place the compression tester back firmly in the plug hole.
- Pull the starter rope at the same pace as before, so that the crankshaft rotates four or five times.
- Whilst cranking watch the needle on the gauge.

If the reading stays about the same as the low "dry" reading then the valves or head gasket may be leaking. If the reading goes up a reasonable amount, the rings and cylinder may be worn and leaking pressure.

The valves and rings are the main places that cause compression loss, but there are other less common places.

Carbon build up in a combustion chamber can result in excessive compression pressure. This can lead to overloading of the engine due to higher combustion pressures and can result in detonation, knocking, overheating, mechanical damage or a blown head gasket.

Possible Causes for Loss of Compression

The illustration (page 112) shows several mechanical problems that cause compression leakage and the trouble that results.

Common symptoms of low compression:
- engine not starting
- engine being hard to start
- engine running erratically

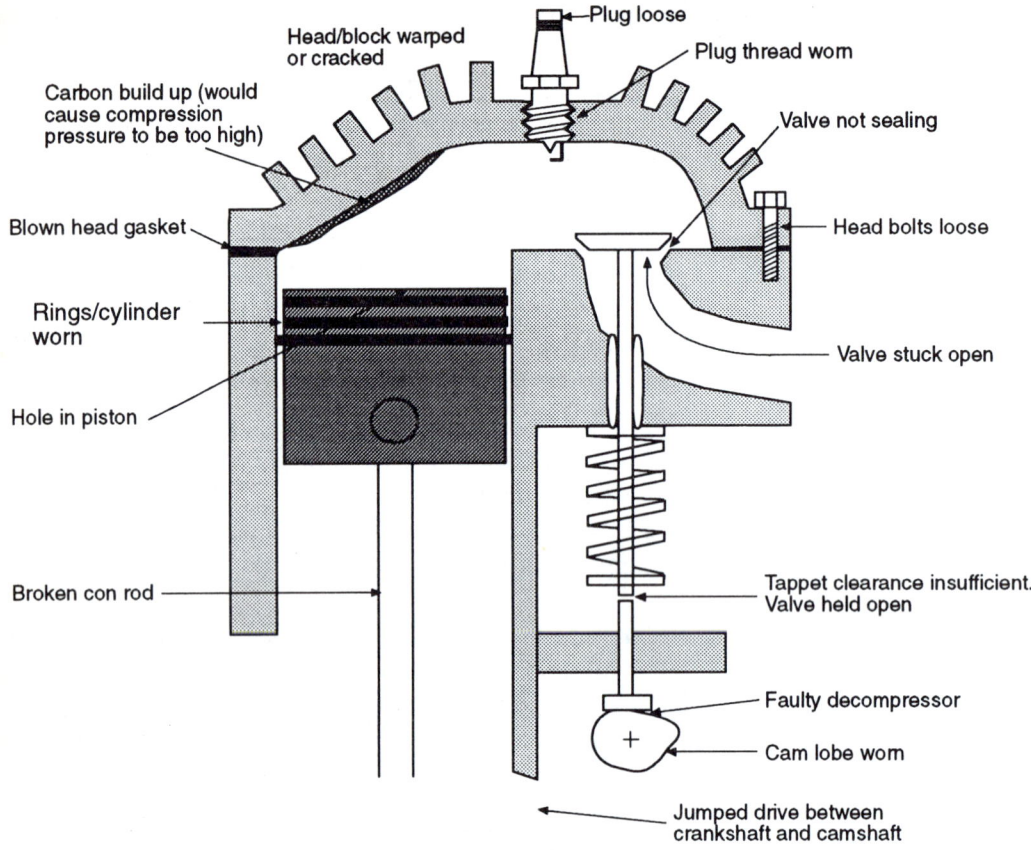

Possible mechanical causes of poor compression

- engine lacking power
- engine misfiring under load
- engine overheating.

If the compression check has indicated that the compression is not adequate, then you have to look for the cause in a step-by-step manner, checking each possible cause out fully before moving on to the next one. You then have to find a remedy to fix the cause. Not only do you have to fix the problem, you have to prevent it from happening again.

Valve Stuck Open

This is a very common problem if the engine is not used every few weeks. A valve sticks open because the engine stops with it open and the stem rusts or gum forms on it. To treat this the head may have to be removed to free the valve.

If not using the engine for a while, leave valves closed. (See Engine Storage, Chapter 15.)

Valve Not Sealing

This can be caused by:
- a burnt valve face
- a burnt valve seat
- carbon between seat and face
- a worn stem or guide
- a bent stem
- a broken or weak spring
- insufficient tappet clearance
- a retainer missing.

All of these should be inspected and repaired as required.

Blown Head Gasket

Caused by:
- eroded or corroded gasket
- loose or broken head bolts
- warped head
- warped cylinder surface
- detonation
- excessive carbon build up increasing compression pressure.

The head gasket must be replaced. Also inspect bolts, head and cylinder block surfaces and repair if required. Remove carbon from the combustion chamber.

Worn Rings or Cylinder

These may be scored or broken. You should replace the rings and hone the cylinder. It may be necessary to rebore the cylinder and fit a new piston and rings, or replace the whole engine.

Physical Engine Damage

- a cracked head or block
- a piston with a hole in it
- a broken conrod
- worn cam lobes
- a faulty decompressor

should all be replaced or repaired if possible.

The best time to do a compression check on your engines is when they are running well. That way you will have a guide for when they play up. The pull of the starter rope is a reasonable indicator of the engine's compression, as is placing the thumb over the plug hole. If the engine has good compression the pressure will tend to blow the end of the thumb off the plug hole. (Ensure ignition is disabled if placing a hand near ignition wires when rotating the crankshaft; also that engine is not hot.)

If the combustion check has indicated that the compression is satisfactory then check out the ignition system.

Ignition

In older engines the ignition systems have two main weak areas that cause trouble: the points and the spark plugs. With the development of the electronic module as a very reliable device to replace the troublesome points, the ignition system is left with only one main weak area, the spark plug.

The spark plug has to work in a very hot and dirty environment and should be cleaned and replaced on a regular basis. If you start the season with a new plug there is a good chance the ignition system will not give you any trouble. It is usually recommended that the spark plug be cleaned or replaced every 100 hours or once a season. If you have an ignition system fault, just simply screwing in a new spark plug may get you mobile again.

The condition of the spark plug is also a good indicator of how well the engine is running. If the engine is worn and using oil, the plug will be smothered with oil. If the fuel mixture is too rich or the spark too weak, the plug will have a coating of dry soot. (See inside back cover.)

The ignition system needs to have all its connections clean and tight. It only takes one dirty or loose connection to cause a problem. Connections get dirty from oxidation, corrosion and rust and the engine vibration shakes things loose.

When you clean the connections do not use water. Use a wire brush and fine emery paper or steel wool.

Checking for Spark

When checking the ignition system, look for anything obvious, such as a broken wire or one that is disconnected, loose or corroded. Make sure the ignition switch is on.

Remove the spark plug for an examination and a clean and re-gap if necessary. If the plug is obviously faulty, fit a new spark plug and start engine.

To check if the plug sparks, place the plug lead onto the plug and lay the plug metal base onto the head metal to obtain a good earth. This should be away from the carburettor and the plug hole where a fire or explosion could occur.

When holding the plug to earth you may be able to rest the plug on the surface or use an insulated screwdriver to hold it. Be careful testing the plug this way because you may get a high voltage shock. If there is such a possibility do not test for spark this way.

Then operate the starter and check at the plug gap for a blue spark. If the plug does not produce a spark, replace it with a new or an identical plug from a running engine and repeat the check. If you get a good blue spark you can then assume the original plug was at fault. So screw in the new plug and the engine could start.

If you do get a blue spark you have established that there is a spark when the plug is out of the engine, but you do not know if it will work in the combustion chamber when it is under pressure on the compression stroke or if the spark is timed correctly.

Inside the combustion chamber on the compression stroke it takes about 8000 Volts to jump the gap at the spark plug which is about 0.75 mm. But producing a spark by earthing the plug on the head outside of the combustion chamber requires less than 8000 Volts.

It takes approximately 10 000 Volts to produce a spark across 4 mm when the gap is not subjected to any compression pressure outside the combustion chamber. So to really test for a good spark, you need a gap of approximately 4 mm. For this you need a special tester or a very steady hand to hold the plug lead, or a metal probe plugged into the lead, 4 mm off the metal of the head. However, it is unwise to test for spark by holding the terminal just off the metal of the head. If you make the checking gap too large between the terminal and the head, say in excess of 12 mm, the spark will not jump that gap and the high voltage generated may damage the coil or the electronic module.

Out on a farm you are unlikely to have a special spark tester. You can purchase spark testers from dealers or you can make a tester using a new non-resister spark plug and bending the earth leg to make a gap of 4 mm.

Connect the special tester, or the one that you have made up, as shown. Make sure the alligator clip makes good metal to metal contact with the head metal.

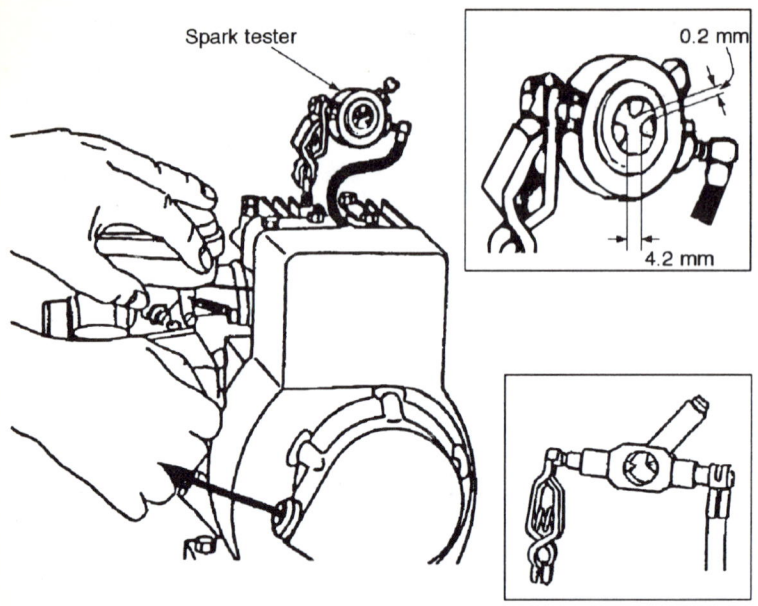

CAUTION: Remember to leave another spark plug in
the head whilst testing for a spark, because petrol fumes
come out of the plug hole and a fire or an explosion could
occur.

With the tester in position, operate the starter and observe
the spark gap in the tester. On engines equipped with electronic modules the crankshaft must rotate at a minimum of
350 rpm for testing. If the spark jumps the gap you can assume the ignition is working.

This tester can also be used to check for spark miss when the
engine is running.

If the engine runs but misses during operation, a quick check
to determine if ignition is not at fault can be made by inserting the tester between the ignition high tension terminal and
the spark plug. A spark miss will be readily apparent because
if the ignition system is faulty the spark will not have enough
energy to jump the gap in the tester and the spark plug.

Flywheel Location Key

If the flywheel key has partially or fully sheared off you could
get a good spark, but the engine may not start or may be
hard to start. This is because the flywheel magnet is timed to
the crankshaft to ensure the spark occurs at the right time.

As was mentioned in Chapter 6, ignition timing of 25° before T.D.C. is fixed on many engines. Sometimes electronic modules can vary this timing by retarding it for ease of starting, then advancing it as the engine reaches operating speed.

Check the flywheel key if you have an ignition problem because the key may be partly sheared. This will cause the timing to be out enough to result in hard starting. With a partially or fully sheared key you will get a good spark with an electronic ignition but it may be hard to start, not start at all and run rough. A points actuated ignition with a sheared key will give a weak or no spark.

Possible symptoms of a weak or no spark
- Engine not starting
- engine being hard to start
- engine running erratically
- engine lacking power
- engine misfiring under load
- engine overheating
- engine knocking.

The spark plug is earthed, in order, to the:
- head
- head bolts
- cylinder blocks
- armature and bolts.

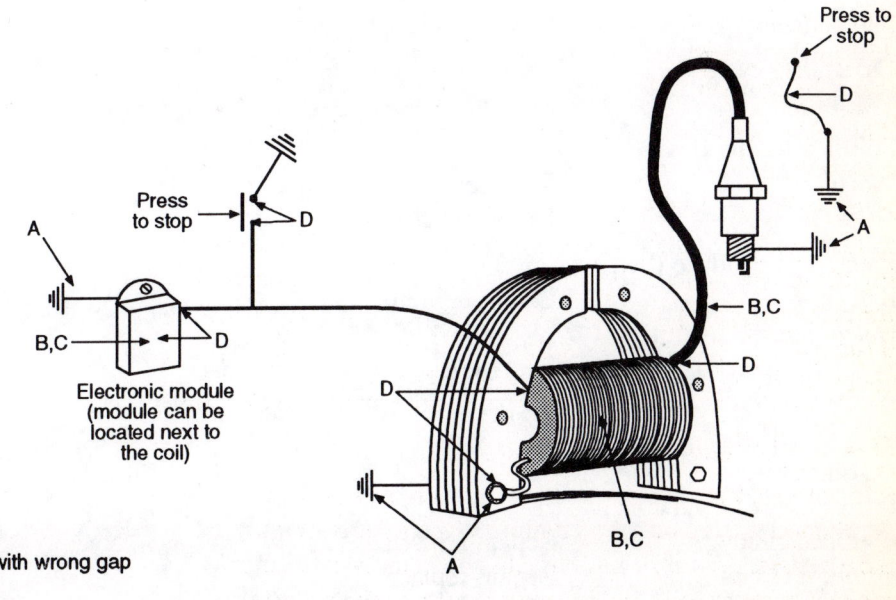

A Bad earth
B Open circuit
C Short circuit
D Bad connections
Magneto ignition
E Points: Dirty, burned with wrong gap
With points
F Points actuator worn

The electronic module on many late model engines is now part of the armature coil. On earlier engines the module was removable. On engines with added modules it is usually attached to the cylinder block.

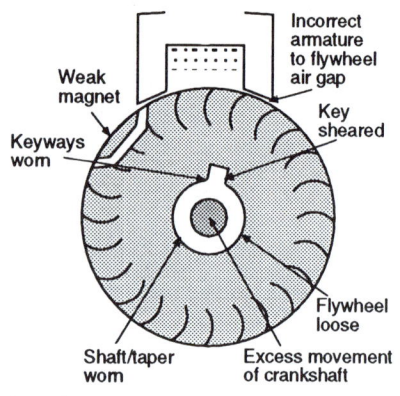

Possible causes for weak or no spark from flywheel area.

Causes and Remedies for Weak Spark or no Spark

Possible causes	*Possible remedies*
Spark plug	
Fouled	Clean, replace
Carbon whiskered	Clean, replace
Insulator cracked	Replace
Wrong gap	Re-gap
Wrong plug	Replace
Ignition switch	
Off	Adjust
Bad connections	Replace, clean
Points	
Dirty	Clean, replace
Burned	Clean, replace
Bad earth	Clean, tighten
Bad connections	Clean, tighten, replace
Wrong gap	Reset
Condenser	
Failed	Replace
Bad earth	Clean, tighten
Bad connections	Clean, tighten, replace, repair
Operating cam	
Worn	Repair, replace
Points opening device	
Worn	Replace
Stuck	Clean, replace
Electronic module (Sealed units)	
Bad earth	Clean, tighten
Failed (internal) from:	
• Bad connections	Replace (some electronic units are part of the armature coil)
• Open circuits	
• Short circuits	
• Burnt out	
Primary wires from coil:	
• Open circuit	Repair, replace
• Short circuit	Repair, replace
• Bad connections	Clean, tighten, repair, replace

Secondary wire from coil:
- Open circuit   Repair replace
- Short circuit   Repair, replace
- Bad connections   Clean, tighten, repair, replace

Armature coil

Failed	Replace if fault is internal
Primary winding ⎤	(Most armature coils come
• Open circuit ⎟	complete with electronic
• Short circuit ⎟	module and primary and
Secondary winding ⎥	secondary wires and are
• Open circuit ⎟	repairable if fault is external
• Short circuit ⎦	of the sealed coil.)
Bad earth	Clean, tighten
Bad external connection	Some repairable
Incorrect air gap to flywheel	Reset

Flywheel

Key sheared	Replace key
Loose	Tighten, replace key
Excessive movement due to worn crank/bearings	Repair, replace bearings
Weak magnet	Replace flywheel
Keyway damaged	Repair/replace flywheel
Taper/shaft worn	Repair, replace

Crankshaft

Keyway damaged	Repair/replace crankshaft

Fuel System

Having checked out the compression and the spark, you then proceed to check out the fuel system.

Do the obvious checks first:
- Is there a good supply of fresh clean petrol in the fuel tank?
- Is the vent hole in the fuel cap clear?
- If you have a gravity feed system is the tap fully on?
- Are the choke and other controls working?
- Is the air cleaner clean and fitted correctly?
- Are the governor and linkages in place?
- Is the carburettor complete and firmly attached to the engine?
- Are the needle valves in the carburettor correctly adjusted?
- Do all the bits and pieces appear to be there and in their correct place?

As was mentioned on page 114 the condition of the spark plug can indicate how the conditions are inside the combustion chamber.

If you are trying to start an engine that has good compression and spark and it will not fire or start, a simple check to determine if petrol is getting to the combustion chamber is to remove the spark plug and examine it.

If the plug is wet with petrol it indicates that petrol is getting to the combustion chamber. If the plug is dry it indicates that petrol is not reaching the combustion chamber. Place a few drops of fresh petrol into the combustion chamber through the spark plug hole. Ensure both valves are closed before putting petrol in. Then replace the plug and lead and operate the starter. If the engine fires a few times then stops it usually indicates that fuel is not getting to the combustion chamber from the carburettor.

CAUTION: when handling petrol like this:
- Do not do it when the engine is hot.
- The ignition must be switched OFF.
- Do not spill petrol on the engine or yourself.
- If petrol is spilt, clean it up properly and wash hands and remove contaminated clothing to avoid fire risk.
- If petrol is spilt, allow time for it to fully evaporate before attempting to start the engine.
- Place any petrol containers well away from engine before attempting restart.
- Do not work in an area where there is a naked flame.
- Do not smoke when near petrol.
- Be aware of all the risks associated with handling petrol.

To see if a good supply of petrol is reaching the carburettor, disconnect the fuel line into the carburettor and see how much petrol runs out the disconnected fuel line.

To determine if there is a good flow of petrol out of the pipe, place a measuring container under the pipe outlet and see how much petrol is in the container after 30 seconds. With a full tank of petrol you should have at least 250 mL in the container after 30 seconds.

The flow rate should also be constant.

If the flow is not good, repeat the check with the fuel cap removed.

If the flow improves, the vent hole in the cap may be blocked. The vent hole is as important as the fuel outlet hole in the tank. As the fuel is removed from the tank the space occupied by the fuel must be replaced by air or the fuel will cease to leave the tank.

After this check you should know in which half of the fuel system the problem lies.

If the fuel flow rate out of the disconnected line appears constant and free then look for the cause of the trouble at any point from where the fuel line fits into the carburettor, to the inlet valve in the engine, plus the governor. If the fuel flow rate out of the disconnected line appears poor, then check all the possible areas from the disconnected fuel line end up to the fuel cap. Check this section first.

Check to see if the vent hole in the cap is blocked. The cap can be cleaned in a suitable solvent and air pressure can be used to clean the vent. Gravity feed fuel tanks are fitted with a fuel tap and some type of filter screen or strainer. The filter screen or strainer may be located in the fuel tank or externally above the filter bowl.

Some engines also use inline fuel filters such as petrol vehicles use.

The most common cause of a restricted fuel flow is a partially blocked fuel filter. The job of the fuel filter is to remove the foreign matter from the petrol so that this foreign matter does not block the jets and passages in the carburettor.

Fuel filters are designed to block up and protect the carburettor if they are not serviced correctly. The fuel filter devices should be cleaned or replaced as recommended, which is usually every 100 hours of service or once a year at the start of the season or more often if required.

If your filter device is blocking up more often you may be using dirty fuel or the tank may need a good clean out.

If the filter can not be cleaned, throw it away and fit a new one. If the filter can be cleaned, such as the filter screens above the filter bowls and the strainers inside the tanks, the devices can be removed and cleaned using suitable solvents and compressed air. Some tanks may require draining before the filter devices can be removed.

Filter bowls usually have taps fitted so that the fuel can be turned off and the bowl then removed.

The filter bowl is a good idea because it allows dirt or water to settle to the lower half of the bowl as the fuel leaves the tank.

If the filter bowl is glass you can keep an eye on the bowl and clean it out accordingly. If you leave it too long the bowl will fill up with water and dirt and block the filter screen and the engine will play up and stop.

Clean and replace the filter devices as recommended by the engine manufacturers.

If the fuel tank has dirt, rust or water in it, it should be removed, cleaned out and refitted. If it has flaking rust in it, the tank may have to be replaced because the flakes of rust tend to block off the outlet hole. The tank is also subject to water from condensation. To reduce condensation the engine should have the tank full of petrol or totally empty and the engine should be kept in a dry area.

The fuel lines should be cleaned in a suitable solvent and blown through with compressed air.

With the vent hole clear, a clean tank, filter devices clean, tap fully on and fuel lines clean, the petrol should have a good run to the carburettor. The next area to check is from the fuel inlet of the carburettor up to the inlet valve on the engine, and the governor.

Some carburettors have a small fuel filter screen on the fuel inlet. If it is easily accessible remove the filter, check, clean or replace it. Then refit the fuel line to the carburettor. Before touching any adjustments on the carburettor, you need to see if petrol is in the fuel bowl. Some carburettors have a drain hole at the base of the fuel bowl. If one is fitted remove the drain plug and see how much fuel runs out the hole with the fuel tap fully opened. Initially there would be a rush of petrol then a steady stream, a bit less than the flow that came out of the disconnected fuel line.

Examine the petrol for dirt or water.

If the carburettor does not have a drain hole, you may have to remove the bowl or remove the high speed mixture needle valve assembly from the fuel bowl to check if the petrol is running into the bowl.

Removing the bowl may only entail undoing one bolt or it might be held in by the high speed needle adjusting valve assembly. If so, gently turn in the valve (clockwise) until it just bottoms out, counting the turns. By doing this you can recheck the setting when you put it back together. With the bowl off and the tap on, the float should hang down and petrol should come out at the needle and seat area and flow into the container at a pace slightly less than that from the disconnected line.

To check the needle and seat for sealing and the float height, you can lightly hold up the float with the end of one finger and the needle should seal off. Usually the float is parallel to the body of the carburettor. If the needle and seat are still leaking with the float lightly held, there may be a fault there.

If the fuel bowl is not removable in a hurry, you may have to remove the high speed needle valve assembly and see how much petrol runs out that hole.

Before removing the needle valve, check its adjustment by counting the turns. With the valve assembly out you should get a steady flow of petrol, governed by the hole for the needle valve and less than that out of the disconnected pipe into the carburettor.

If fuel does not run out into the carburettor bowl the usual cause is a sticky needle or a blockage in the passage above the needle tip.

The needle can be freed up and cleaned or it can be replaced as a unit or as a separate item. Other causes could be that it is blocked by dirt or insects have built a home there.

Most needles are attached to the float arm with a spring, so that the needle is pulled open as the float drops. The spring sometimes serves as a shock absorber.

If the petrol does not seal off at the needle and seat when the float is lightly held, the needle and/or seat may need replacing or a good clean.

If the float height is incorrect it is usually corrected by bending the tong on the float. Do not bend the float to adjust. Always check with the manufacturer for the correct setting and methods of checking and adjusting.

Put the carburettor together again and establish that petrol is in the fuel bowl. If the needle and seat leaks now, you would notice petrol running out of the carburettor and further checking is then needed to find out why.

If the engine is still not starting or running, we continue the search for the problem.

It is recommended to check the base or initial settings of the idle and high speed mixture needle adjusting valves if they are fitted. Most carburettors are now using fixed main jets instead of the adjustable high speed valve, but nearly all carburettors still use an adjustable idle mixture valve.

The valves can be turned lightly to check how far they turn in until they bottom out. Note the readings. They should only turn in one to two and a half turns on average. It also pays to take the valves out if this is not a major problem and look at the tips to see if they are clear. Some valves are hollow, so check out the small holes in them.

Screw the valves in fully, then turn out 1 1/2 turns each on average, unless you have the manufacturer's specification. The engine should start on this setting, then it can be correctly adjusted with the engine running. The idle speed screw should not need adjusting.

After checking and setting the needle valves, then see if the engine will start and run. If it does not you continue the search.

Before getting deeper into the carburettor, look for faults that could exist, such as loose screws, nuts or bolts that hold the carburettor together, that you may have overlooked during your first check over of the carburettor at the start of the troubleshooting exercise. Also, look for leaking gaskets where parts are bolted together, such as where the carburettor bolts onto the engine. If it sucks in air at a gasket, it can usually be detected by using oil. Squirt oil onto the join and see if it is drawn in as you attempt to start the engine or as it runs.

Also run your eye over the governor linkages and spring again. The engine usually starts with a faulty governor but will not control its speed very well. Check your controls again such as the throttle and choke and make sure they work. If they all check out, that leaves the bowels of the carburettor.

As we know from Chapter 5, carburettors have many small holes and passages that could easily be blocked by dirty petrol, gum or varnish. This would upset the carburettor so that the petrol and air will not flow as they should and the engine will then not start or run properly.

The most common thing to block are the jets or holes in the nozzles. Most carburettors are designed so that the nozzles and jets are easily removable for cleaning without the need to pull the carburettor off the engine and dismantle it. To pull the nozzles or jets out you need properly fitting tools and screwdrivers. Bad screwdriver tips will do a lot of damage.

Examine the holes in the jets and nozzles using good light, to see if you can see a fault such as a partial or fully blocked hole. Wash the parts in suitable solvent and blow out with compressed air and look again at the hole sizes.

Be careful cleaning out blocked holes. The holes are precision drilled and if you dig away with an object and alter the hole size, the part may never work the same again.

Before refitting the jets and nozzles you may be able to place the air compressor hand nozzle near the hole for the jets and nozzles and give them a blast of air to try and encourage any dirt to move.

Refit the jets and nozzles carefully. Most screw straight in and bottom out. Refit adjustable needle valves and set them to a base or initial setting.

Turn the fuel tap on and see if the engine will start and run correctly. If it will, adjust the carburettor to manufacturer's specifications or as was discussed in Chapter 5.

If the engine fails to start and run correctly, the carburettor may require a pull down and check out. You can do it yourself if you are proficient and have the correct tools and instructions, or you can take it to a dealer.

Suction Feed Carburettors

On these types of carburettors the correct operation of the choke is important to start the engine when it is cold. These carburettors may use an automatic or manual choke. If the choke is not working correctly, the engine has difficulty starting when cold; the automatic choke also plays a role in enriching the mixture on acceleration.

On the base of the fuel pickup tubes there is a fine mesh fuel filter screen. Cleaning this screen entails the removal of the carburettor from the tank.

On the type of carburettor that draws the petrol out of the tank directly to the needle valve, there is a check ball in the tube, like a foot valve in a water pump, to hold the petrol in the tube and stop it dropping back into the tank. This ball must move freely. If this check ball jams closed, fuel will not be able to go up the tube. If it jams open, fuel will not be held in the tube, causing starvation.

The needle valve can be removed for inspection. Replace the needle valve if it is bent, grooved or broken. Check the condition of the needle seat and replace it if damaged. The two metering holes can be cleaned with compressed air.

On the type of carburettor that pumps the petrol for the tank into a fuel cup there are two pickup tubes with mesh screens at their base. There are no check balls used, for the valves in the diaphragm pump hold the petrol in the pickup pipe as it is being pumped into the fuel cup. Within the diaphragm pump the valves could get dirt under them so that the pump does not work, and the diaphragm could perish and split, thus causing the pump to malfunction.

(Please refer to pages 128–9 for a troubleshooting chart on the fuel system.)

Other Engine Problems

Engine Overheats

Causes	*Remedies*
Oil level too low or too high in crankcase	Set to correct level
Fins missing on flywheel	Replace flywheel
Air cooling fins blocked	Clean fins
Screen on blower housing blocked	Clean screen
Shrouds or deflectors not in place or missing	Fit correctly
Incorrect spark plug	Fit correct spark plug
Ignition timing incorrect	Re-time
Carburettor set too lean	Adjust to a slightly richer mixture

Engine Knocks/Rattles

Causes	*Remedies*
Oil level low in crankcase	Set to correct level
Wrong oil in engine	Drain crankcase and refill with correct oil
Oil diluted	Drain crankcase and refill with correct oil
Excess running clearance inside engine: • piston to cylinder	Engine to be dismantled for checking

Engine Knocks/Rattles (continued)

Causes	*Remedies*
• piston pin to piston • piston pin to small end hole of con rod • big end of con rod to big end journal • main bearings	Engine to be dismantled for checking
Broken parts in engine	Engine to be dismantled for checking
Loose flywheel	Tighten/replace
Engine overloaded	Operate correctly
Piston or valves hitting carbon deposit	Decoke combustion chamber

Engine Using Excessive Oil

Causes	*Remedies*
Faulty breather	Repair/replace
Worn or broken piston rings	Check parts, repair/replace
Worn cylinder bore	Check parts, repair/replace
Worn valve stems/guides	Check parts, repair/replace
Valve stem oil seals worn/faulty	Replace seals
Incorrect oil used	Use correct oil
Gaskets, seals leaking	Repair/replace

Troubleshooting Two Stroke Engines

Troubleshooting of a two stroke engine is similar in many ways to that carried out on the four stroke engines.

These differences have to be kept in mind as you troubleshoot:
• use of a dry sump
• importance of the crankshaft seals
• use of ports and reed valves
• use of petrol and oil mixture
• use of diaphragm carburettors
• governors are not normally used
• carbon builds up in the exhaust port and muffler.

Troubleshooting chart on the fuel system

Area	*Possible trouble*	*Possible causes*	*Possible remedies*
Air vent	Restricted petrol flow out of tank	Vent holes blocked by dirt	Unblock holes
Fuel cap	Petrol contaminated	Cap loose/worn Gasket missing/damaged	Tighten/replace cap Replace gasket
Fuel	Tank empty	Used up, evaporated, stale, contaminated with water etc.	Fill tank with fresh clean petrol
	Does not ignite correctly	Dirt, water etc. in petrol	Replace with fresh clean petrol. Clean out tank
	Rust, scale	Water contamination	Clean, replace tank
	Gum, varnish deposits	Petrol evaporates whilst engine in storage	Correct storage required
	Dirt, water	Contaminated petrol	Clean out tank
Fuel tap	Fuel flow restricted Blockages	Tap closed or partly closed	Open tap fully
		Contaminated petrol dirt, water, etc.	Clean out
		Gum, varnish deposits, rust, scale from tank	Clean fresh petrol
Fuel tank outlet	Blockages	Contaminated petrol	Clean out
Fuel filter devices	Blockages	Contaminated petrol	Clean out
Fuel lines	Blockages	Kinked Contaminated petrol	Repair, replace Clean out
Needle and seat	Blockages Flooding Runs excessively rich Starving	Contaminated petrol Needle stuck open Needle/seat worn Needle stuck closed	Free up, replace Clean Replace Free up, replace
Float	Runs rich	Float set too high Hole in float or porous	Set correctly Repair/replace
	Runs lean	Float set too low	Set correctly
Needle valves			
• Idle	Runs rich at idle Runs lean Will not idle correctly	Incorrectly adjusted Incorrectly adjusted Worn/damaged valve	Correct the adjustment Correct the adjustment Replace valve
• High speed	Runs rich at speed Runs lean	Incorrectly adjusted Incorrectly adjusted	Correct the adjustment Correct the adjustment

Troubleshooting chart on the fuel system (continued)

Area	Possible trouble	Possible causes	Possible remedies
Needle valves (cont)	Will not accelerate correctly	Incorrectly adjusted	Correct the adjustment
	Will not run smoothly	Worn/damaged valve	Replace valve
Jets, nozzles, bleed holes	Blockages	Contaminated petrol	Clean out
		Dirt entering from carburettor throat	Service air cleaner. Locate leaks or worn or loose parts
	Runs rich	Parts loose, incorrectly fitted, damaged	Tighten, repair replace
	Runs lean	Holes closed up	Clean out
Throttle, valve and shaft	Sucks air, dirt etc around shaft	Worn shaft and hole	Repair, replace
	Throttle valve comes adrift	Vibration	Repair, replace
Idle speed screw	Falls out/moves	Vibration	Replace/adjust
Choke valve	Not fully closing	Incorrectly adjusted	Adjust correctly
	Not fully opening	Incorrectly adjusted	Adjust correctly
		Faulty controls	Repair, replace
Governor	Mechanical governor disintegrates	Excess centrifugal force, low oil level	Repair, replace Maintain oil level
	Spring, parts, fail, loosen, wear out, bend	Vibration or rough handling	Repair, replace adjust
Throttle control	Out of adjustment	Vibration	Adjust correctly
Air cleaner	Partial blockage	Dirt, dust	Clean
	Comes loose	Vibration	Tighten
	Holes in element	Wear and tear	Replace element
Gasket	Sucks in air, dirt etc.	Loose bolts/nuts Broken gasket Vibration	Tighten, replace
Passages, drillings	Blockages	Dirt entering from carburettor throat	Service air cleaner Locate loose or worn parts
		Contaminated petrol Dirt, water etc. Gum, varnish deposits Rust, scale from tank	Clean out Correct storage Clean fresh petrol

Powered Equipment Affecting Engine Operation

Often what appears to be an engine problem may be the fault of the equipment that the engine powers rather than the engine itself.

Some equipment can not be disconnected easily from the engine, such as generators and water pumps, so if a fault exists, always keep in mind that it could be a seized bearing or something similar in the equipment. This makes the engine hard to start and can stop it developing full power.

The following are some of the conditions that could exist due to a fault in the equipment that is powered by the engine, rather than in the engine itself.

Power Loss

Causes	Remedies
Bind or drag in powered equipment	If possible, disengage the engine and operate equipment by hand to feel for binding action
Material building up on powered equipment to slow it down	Clean away material
No lubrication in transmission	Check for damage Fill with oil
Over-tensioned drive belts	Tension correctly

Noise

Causes	Remedies
Worn couplings	Replace
Bearings worn	Replace
Little or no lubrication in transmission	Check for damage Fill with oil

Hard Starting, Kickback or Failure to Start

Causes	*Remedies*
Starting under load	Ensure load is disengaged Lighten load if possible
Loose belt (gives backlash effect)	Tension belt correctly
Interconnecting controls, wires and switches	Check for serviceability
Loose blade or base plate (gives kickback effect)	Tighten. Check for partly sheared flywheel key

Vibration

Causes	*Remedies*
Mounting bolts loose	Tighten
Incorrectly mounted or mounting cracked	Remount correctly/repair
Worn couplings	Replace
Crankshaft bent/twisted	Replace
Cutter blade/base plate bent or out of balance	Remove and balance check for partly sheared flywheel key

Chapter 14

Engine Selection and Application

When you buy a piece of equipment that is powered by a small petrol engine, the engine should be capable of supplying sufficient power to that piece of equipment to suit all normal operating conditions.

If you are replacing an old engine you should buy the same horsepower engine again. If you replace it with a smaller engine, the equipment may not work as it should and the engine will be overloaded and cause you some problems.

If you replace it with a more powerful engine, it will do the job more easily but the higher power may overload the powered equipment and cause damage. A more powerful engine does not always prove to be an economical proposition.

When you go to buy a tractor the dealer will show you graph charts that indicate such things as the power and torque of the engine at different operating speeds. He will also tell you at what speed to best operate the tractor's engine. When you go to buy a small petrol engine, whether it be attached to a piece of powered equipment or not, the dealer should be able to show you similar graph charts.

When engine manufacturers build engines they test them with a machine called a dynamometer to determine torque and power ratings at various speeds under full load.

The engines are usually equipped with all the bits and pieces that are sold with the engine, such as air cleaner and muffler, when they are tested. The engines are run at full throttle at operating temperature and a brake is applied to the crankshaft to slow the engine down. This will give a torque reaction which is measured at various revolutions per minute.

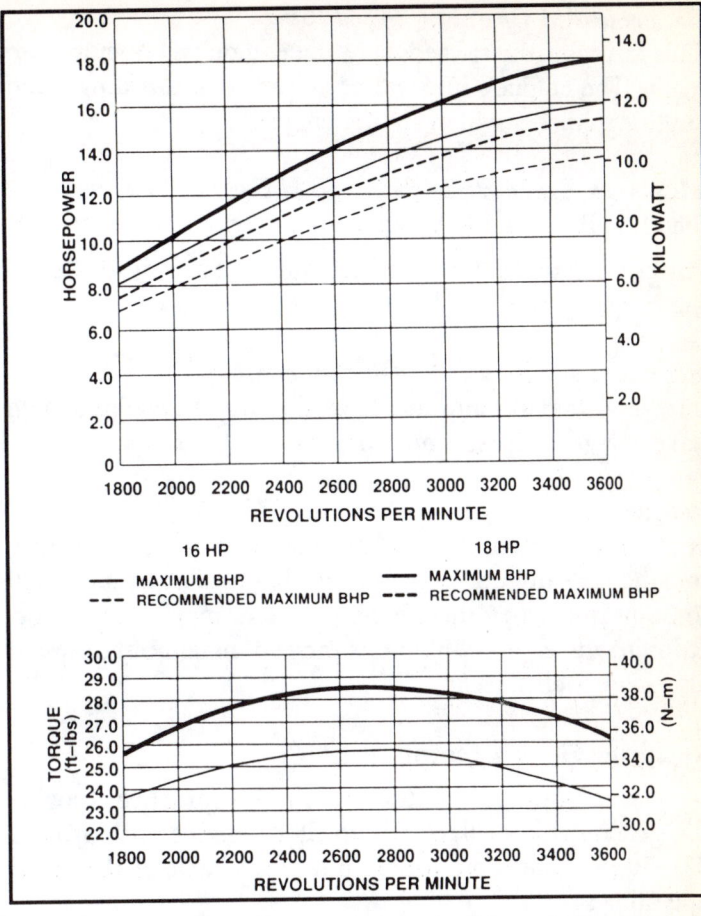

Horsepower and torque curves

These results are then recorded on a graph chart, the type which the dealer can show you.

Tests are conducted under standard conditions of air pressure and temperature. Therefore, if the engines are used at elevations above sea level and higher temperatures where the air is lighter the engine power will decrease. For practical reasons, it is recommended that for continuous operation the engine horsepower loading should not exceed 85 per cent of the rating.

If an engine is directly coupled to a piece of equipment, that equipment receives the same power as the engine produces. However, if an engine uses a power transfer method such as a reduction gearbox or a V-belt and pulley drive to power a piston water pump, you are going to have power losses of about ten per cent in the gears and belt drive. The belt must also be correctly tensioned to reduce power losses.

Recommended Operating Speed Range

This indicates the range where the engine will operate normally. The engine's idle will be slower than this range, and anything quicker will strain the engine.

Maximum Brake Horse Power (B.H.P.)

The B in B.H.P. is the brake from the dynamometer testing.

Power is relative to torque of the engine at speed, so altering the torque or speed affects the power accordingly.

Recommended Maximum Operating B.H.P.

This is the line that indicates at what power the engine should have at a given speed under normal conditions.

Torque

As we mentioned earlier, this line indicates how well the engine breathes. It can be seen that it breathes well at around 2500 to 3000 rpm when the engine produces maximum torque. Running the engine within this range of maximum torque is also more fuel efficient.

Replacing Electric Motors

If you are replacing an electric motor with a small petrol engine the general rule of thumb is to fit an engine with around 25 –50 per cent more usable power available at the given operating speed.

Operating Points

If you have built something that you want to use a small petrol engine to power, you should look around at similar types of equipment and get opinions from various people, including engine dealers, as to how much power is needed to operate it.

Engine dealers also have information available such as mounting bolt patterns and the size of drive shafts and key types used.

Chapter 15

Engine Storage

When an engine has done its job for the season, take a few minutes to store it correctly and it will stand a better chance to start up and run for you the next time. If you do not store it correctly there is a good chance you will get plenty of troubleshooting experience in the following season.

An engine should be stored if it is not to be used within 30 days from the last start up. The following steps should be undertaken.

1. Clean the Engine

Remove the blower housing cover and clean the engine using compressed air and suitable solvents. Do not use solvents on the ignition system.

Avoid the use of water, because it gets into the fuel and ignition systems and causes trouble later on.

While the cover is off check over the condition of the engine to see if everything is in its place and secure and if any repairs are needed.

2. Repairs

Carry out any necessary repairs. Service the air cleaner. Note its condition in case it needs replacing.

3. Engine Oil

While the engine is hot, drain the oil and refill. Also do any other oils, such as the reduction gearbox, if fitted.

4. Petrol

Engines should not be stored with petrol in the fuel system because it forms gum and varnish deposits which can block the fuel line, jets, passages and damage diaphragms. Petrol only has a life of about three months before it goes stale.

Drain the fuel tank and run the engine till it uses the petrol out of the carburettor and stops. To get the remaining drops of petrol out of the carburettor, a drain plug is sometimes fitted.

Some engine manufacturers recommend the use of a special fuel additive that enables the petrol to be left in the fuel system for an extended period and keeps the petrol fresh.

5. Oiling the Cylinder

To prevent the formation of rust in the top of the cylinder and around the valves, they have to receive a liberal coating of oil. Remove the spark plug and insert an oil can spout, containing engine oil, in the plug hole and give it a few pumps.

Work the starter a few times to circulate the oil. Replace the spark plug and work the starter a few more times to encourage the oil onto the valve stems. Then set the piston at T.D.C. on the compression stroke. This way the piston is at the top of its travel to cover the cylinder and both valves are closed to cover the working part of the valve stems in the valve guides.

If a valve is left open, particularly the exhaust valve, the stem tends to rust, causing a sticky valve next season. Having both valves closed also protects the cylinders from foreign matter.

6. Store in a Suitable Area

The engine should then be covered and stored in a clean, dry, well ventilated area away from animals.

If there is a vermin problem take the necessary precautions to stop them damaging the stored engine.

Preparing the Engine for Work Once it Comes out of Storage

Before the engine can be put to work it has to be properly prepared.

1. Clean the Engine

Remove the blower housing cover and check for vermin or insect nests. Use compressed air to thoroughly clean the cooling fins, governor controls and so on. Refit the blower housing and with the ignition off, check that the engine is free to rotate and that it appears to have compression.

2. Air Cleaner

Service the air cleaner. If fitted with a replaceable element, it should be replaced at the start of each season.

3. Petrol

Fill the tank with fresh clean petrol.

4. Spark Plug

Fit a new spark plug. The old plug has probably fired a few million times, so retire it. A new spark plug is only a few dollars and is often a cheap tune up.

5. Oil

Check all oil levels.

Start the engine and warm it up.

When warm, stop the engine and drain out the oil because it was probably contaminated by condensation during the storage period. Refill it to the required level with the correct oil.

6. Final Check and Adjust

Look over the engine and check that everything is in place and secure. Start the engine and listen for odd noises.

The carburettor may require slight tuning.

If all appears well, the engine is ready for another season.

Glossary

A

Air bleed A small hole that allows air to bleed into the fuel that is being discharged from the carburettor. Helps to atomise the fuel and correct the fuel-air mixture.

Air cleaner A device through which air is drawn to remove dust, dirt etc. to provide clean air to the engine.

Air cooled An air cooled engine is cooled by air passing over fins on the cylinder head and block. Air flow over fins is usually caused by the fan action of the rotating flywheel.

B

Big end The crankshaft (big) end of a connecing rod.

Bore Refers to the cylinder or to the diameter of the cylinder.

B.D.C. Bottom Dead Centre Is the position of the piston when it is at the bottom of its stroke in the cylinder.

B.H.P. Brake Horse Power An imperial system unit of measuring engine power output. The metric system states power output in kilowatts.

Butterfly A disc on a spindle used to control air flow, such as with a choke or throttle butterfly as used in a carburettor.

C

Camshaft A shaft with cams to open each valve at the correct time. Rotates at half engine speed.

Capacitor Also called a condenser. Reduces arcing at ignition points by absorbing electrical surges when points open.

Carburettor Device used to mix fuel with air in the correct proportions to suit the load and speed requirements of the engine.

Choke A valve located in a carburettor to enrich the fuel-air mixture for starting the engine.

Combustion Process of the burning of the fuel-air mixture within the combustion chamber of the engine.

Combustion chamber Area above the piston with the piston at T.D.C. where the fuel-air mixture is ignited by the spark plug and burnt.

Compression Compressing of the fuel-air mixture charge in the cylinder as the piston travels from B.D.C. to T.D.C. during the compression stroke.

Compression ratio The amount that the fuel-air mixture is compressed in the cylinder. This is the ratio of the volume of the cylinder and the combustion chamber with the piston at B.D.C. compared with the volume with the piston at T.D.C.

Compression stroke The upward travel of the piston to compress the fuel-air mixture. The second stroke of the operating cycle of a four stroke engine with the piston moving up the cylinder and both valves closed.

Connecting rod A metal bar that connects the piston to the crankshaft.

Cycle A series of repetitive actions, such as two and four stroke cycles of an engine.

Cylinder A circular hole in the engine block in which the piston moves up and down.

Cylinder block The main part of the engine that contains the cylinder and usually has a crankcase attached.

Cylinder head Metal section bolted to top of cylinder block. Covers top of cylinder and forms combustion chamber. On O.H.V. engines it also contains the valves.

D

Detonation Fuel-air mixture burning too violently, almost exploding.

Displacement The total volume of air displaced by the piston travelling from B.D.C. to T.D.C. A measurement used to indicate an engines size. Imperial system uses cubic inches (ci) and metric system uses cubic centimetres (cc) or litres (L).

E

Earth Electrical term for ground, a common connection on the negative side of the electrical system.

Electrode Spark plugs have a centre electrode and an earth electrode with a cap between them, across which a spark occurs.

Electronic ignition An ignition system that uses non moving parts such as transistors to do the job of the movable ignition breaker points.

Exhaust gas Gases formed by combustion, which includes the deadly carbon monoxide gas.

Exhaust stroke The movement of the piston to expel the exhaust gases from the cylinder. The fourth and last stroke of the operating cycle of a four stroke engine with the piston moving up the cylinder with the exhaust valve open.

External combustion engine An engine where the combustion of the fuel takes place outside the engine cylinder, eg a steam engine.

F

Flywheel Heavy wheel attached to crankshaft to smooth out firing impulses. Provides inertia to keep crankshaft rotating during periods when no power is being applied.

Four stroke cycle An engine operating on the Otto cycle where two revolutions of the crankshaft, which gives four strokes of a piston, are needed to complete a firing cycle. The four strokes are intake, compression, power and exhaust.

Fuel mixture A combustible mixture of fuel and air. An average fuel mixture, by weight, would be 15 parts of air to one part of petrol.

G

Gasket A sealing material between two surfaces.

Governor A controlling device used to limit the speed (R.P.M.) of an engine.

H

Heat range (spark plugs) Refers to the range of temperatures within which the plug is designed to operate.

High tension The high voltage produced by the ignition coil. Also refers to the secondary wire that runs from the coil to the spark plug.

Hot spot Localised area with a higher than normal temperature.

I

Idle The slow speed that an engine will run at with the throttle closed.

Ignition Firing of the compressed fuel mixture in the combustion chamber by a spark from the spark plug.

Ignition breaker points A movable set of contacts (a switch) that are used to make and break the primary circuit of the ignition system so as to produce a spark at the spark plug.

Ignition coil An induction coil which produces high voltage for ignition.

Ignition timing The time, in relation to the turning of the cranksharft, at which the spark occurs at the spark plug to ignite the fuel-air mixture. Usually occurs just before T.D.C. as the piston nears the end of its compression stroke.

Intake stroke The movement of the piston that draws the fuel-air mixture into the cylinder.

J

Jet A part of a carburettor with a small hole in it so as to control the flow of petrol.

K

Key A metal insert fitted between two parts to prevent them from slipping.

Kilowatt A metric unit for measuring power.

L

Land The metal separating the ring grooves in a piston.

Leaded petrol Petrol containing tetraethyl lead, an antiknock additive.

Lean mixture A fuel mixture with excessive air in relation to fuel.

M

Magneto An engine driven device that generates high voltage for ignition. Does not need a battery to operate it.

Muffler A device in the exhaust system that reduces exhaust noise.

O

O.H.V. Overhead valve engine.

Oil bath air cleaner An engine air cleaner that uses oil to remove dust particles.

Oil consumption The oil used or burnt in an engine.

Oil control ring A piston ring used to control oil on the cylinder wall.

Otto cycle The operating cycle of a four stroke engine.

P

Petrol A hydrocarbon fuel used in internal combustion engines.

Piston A round plug that slides up and down inside a cylinder

Poppet valve Valve with a head and a stem. Such valves are used to open and close the port entrances to the cylinder head.

Power The rate of doing work.

Power stroke The working stroke of an engine during which the piston is forced down the cylinder by the pressure of the expanding gases from combustion.

Pre-ignition Fuel mixture being ignited before the correct time.

Push rod Rods used to transfer movement of a camshaft to actuate valves on an O.H.V. engine.

R

Reach Length of the thread of a spark plug.

Reciprocating motion Backwards and forwards movement, such as a piston in a cylinder.

Rich mixture A fuel mixture with excessive fuel in relation to air.

R.P.M. Revolutions per minute. Refers to speed of a shaft, such as crankshaft speed.

Rotary motion Continual motion in a circular direction, such as that of a crankshaft.

S

Sediment Matter which settles in the bottom of a liquid, such as dust in a fuel bowl or tank.

Small end The piston (small) end of a connecting rod.

Spark plug A device screwed into the combustion chamber that is used to ignite the fuel mixture.

Spark plug gap The gap in the spark plug, between the electrodes, that the spark jumps to ignite the fuel mixture.

Stroke Movement of a piston from T.D.C. to B.D.C. or from B.D.C. to T.D.C.

T

Tappet Part of an engine's valve mechanism.

Throttle butterfly A valve in the carburettor that controls engine speed and power.

T.D.C. Top dead centre. Refers to the piston when it has reached the top of its travel in the cylinder.

Torque Twisting or turning force. Torque is produced by the engine at the crankshaft.

Transistor An electronic switching device used in electronic ignition to replace points.

Tune up Checks and adjustments to an engine to restore performance.

Two stroke Engine that operates on a two stroke operating cycle.

V

Valve train The valve operating mechanism from the camshaft to the valves.

Venturi Part of the carburettor through which air flows to draw fuel into the air flow stream to form the fuel-air mixture when the engine is above idle speed.

Volatility The tendency of a liquid, such as petrol, to vaporise.

Acknowledgements

The author would like to thank the following for their assistance with the preparation of this book:

Anthea Hubbard for typing the manuscript.

The following Cowra businesses for their technical assistance

    Farm Services

    Ian Davidson Machinery

    Ken Dawes Mowers and Chainsaws

    Phil Tomlin Honda

Librarian Marion Mitchell and her colleagues in other centres.

The publisher acknowledges the generosity of both Briggs and Stratton Australia Pty Ltd and Honda Australia for the use of technical drawings, diagrams and information in this book; and Champion Spark Plug Co. for the chart on the inside back cover.

Index